The Multiverse of Consciousness: Unraveling the Illusion of Spacetime

DAVID WOODSON II

"X" - @woodson1900

davidwoodson84@gmail.com

Copyright © 2024 David Woodson

All rights reserved.

CONTENTS

1 THE ANOMALY OF CONSCIOUSNESS   1
2 THE UNIQUENESS OF LIFE IN THE   17
  UNIVERSE
3 THE EVOLUTION OF LIFE AND   36
  CONSCIOUSNESS
4 THE WORLD BEYOND OUR SENSES   45
5 TIME AS AN ILLUSION   56
6 THE NATURE OF SPACE   75
7 MATHEMATICAL REALITY   97

CHAPTER 1: THE ANOMALY OF CONSCIOUSNESS

The Virtual Reality You Call Life

I address you directly, reader, and begin with a perhaps unexpected thought: you live in a virtual reality. Yes, all the sensations you experience from birth to death—the colors of a sunset, the taste of your favorite food, the touch of a loved one—are created by your brain. It constructs a virtual environment within itself that can accurately replicate the external, real environment. How can you be sure that the reality you perceive is genuine? Are there any anomalies that suggest its artificiality?

Think about this for a moment. You sit in a room, reading these lines, perhaps listening to music or smelling coffee. All these sensations are created by electrical impulses in your brain. They become your reality, your truth. But what if these impulses could be altered or even created by an external source?

Consider some examples. You wake up in the middle of the night feeling like you've fallen. Even though you were lying still in your bed, your brain created the illusion of falling, and you actually felt the fear. This is a small example of how the brain can make you believe something that isn't real. But how deep can this illusion go?

Think about the colors of a sunset. They can be incredibly vibrant and beautiful, yet they are simply waves of light that your eyes perceive and transmit to your brain, where these signals are interpreted as colors. How do we know these colors exist outside of our consciousness? They are just wavelengths on a spectrum.

The taste of your favorite food is another example that demonstrates how our brain creates reality. When you eat, your brain receives signals from taste receptors and converts them into the sensation of taste. But what happens if these receptors are tricked? Scientists are already creating devices that can stimulate taste receptors, causing the sensation of different tastes without consuming any food. Thus, the taste of your favorite food is just a product of your brain's activity.

The touch of a loved one is one of the most intimate and important sensations. But even this is just electrical impulses transmitted through nerve endings to the brain. Imagine a technology that could create these impulses artificially. You would feel the touch even if no one was around. Take dreams, for instance—they seem so real while you're asleep, but as soon as you wake up, reality returns. This indicates that our brain is capable of creating entirely convincing illusions. Can you be sure of the authenticity of these sensations?

The Simulation Hypothesis: Are We Living in the Matrix?

One of the most popular ideas that comes to mind for modern people when pondering the nature of consciousness is the simulation hypothesis. The idea that we might be living in a giant computer simulation became especially popular after Nick Bostrom's paper "Are You Living in a Computer Simulation?" He argues that if humanity reaches a sufficient level of technological development, it could create countless simulations of the universe, indistinguishable from the original. Within these simulations, people would emerge who could also create their own simulations, and so on ad infinitum. Therefore, the probability that we are at the beginning of this chain, rather than in one of the countless simulations, approaches zero.

This idea, while seemingly fantastic, has quite serious philosophical and scientific foundations. It makes us think about the nature of reality and how deceptive our senses can be. If we truly live in a simulation, then everything we consider real—the laws of physics, our bodies, other people—might just be part of a complex program.

Think about the possibilities that such a hypothesis opens up. If we live in a simulation, then our world could be changed or restarted at any moment. It also means that there is someone or something controlling this simulation, creating our sensations and experiences. It could be another civilization that has reached an incredible level of development, or even future generations of humanity who decided to recreate the past in the form of a simulation.

There are other aspects of this hypothesis to consider as well. For example, what would happen if we could find evidence that we live in a simulation? Could we interfere with its operation, changing the rules of the game? Or would we remain helpless observers, doomed to a life in a virtual world?

Consciousness: The Anomaly That Breaks the Illusion

But even if we discard the simulation hypothesis, there is at least one anomaly in the world you know that cannot be explained by the known laws of physics. This anomaly is consciousness.

Imagine that your memory was erased and you were placed in a virtual reality with different laws of physics. At first glance, you wouldn't be able to learn anything about the world outside the simulation. But this isn't true. By studying the laws of this new world, you would inevitably conclude that they are too simple to give rise to such a complex phenomenon as consciousness. The very fact of your existence as a sentient being would contradict the laws of this world.

Consciousness is a unique phenomenon that goes beyond the traditional understanding of physics and biology. It includes self-awareness, the ability for abstract thinking, reflection, emotions, and intuition. It's what makes you "you," giving you personality and individuality.

Even in our world, consciousness remains an unsolved mystery. We can explain many aspects of how the brain works, from neural connections to chemical reactions, but we still don't understand how consciousness arises. This anomaly points to the existence of something more than just physical processes.

Now imagine that you are studying the laws of the new world you've been placed in. You notice that these laws are very simple and logical. They can explain many things, but they cannot explain why you have consciousness, why you feel, think, and experience. This would make you wonder if there is something beyond this world, something that goes beyond its laws and limitations.

This could be the key to understanding that even in our real world, there is something that goes beyond the material. Consciousness could be an indication that our reality is not just a physical world, but something much deeper and more complex.

The Chessboard as a Metaphor

Imagine you found yourself in the world of chess, becoming the king piece. After learning the rules of the game, you would realize that they are too primitive to explain your ability to think, feel, and be aware of yourself. You would understand that your existence cannot be explained by the laws of this chess world.

The world of chess is strictly defined: each piece has its own clearly defined rules of movement and limitations. The moves of the pieces are entirely subject to these rules. However, if you were the king and had consciousness, you would immediately see that the rules of the game do not explain your ability to reflect on these rules, to feel emotions from victory or defeat, to be aware of yourself as a separate individual in the chess world.

Similarly, consciousness cannot be explained by the known laws of physics. It is something more than just electrochemical processes in the brain. It is what makes us unique and distinguishes us from all other objects in the universe.

In the chess world, your ability for self-reflection and thinking would go beyond the concepts that describe this world. It would become an anomaly that contradicts the internal laws of the game. By studying the world of chess, you would understand that your thinking and consciousness cannot be the result of only those rules that govern the movements of the pieces.

The same applies to our reality. The known laws of physics successfully explain the behavior of matter and energy, but they cannot explain how consciousness arises. We can study neural connections, chemical processes in the brain, but this does not fully reveal the nature of our consciousness.

Causal Closure of the Physical World: The Illusion of Self-Sufficiency

The more sophisticated the simulation, the longer you will remain in the illusion of the reality of your simulated world. If your "jailers" want you to never guess that you are in a simulation, it's not enough for them to simply shield you from observing the outside world. The simulation must be completely self-sufficient, self-consistent in everything that concerns explanations. The answer to every question you have about the world around you must be woven into the structure of the simulated physical reality.

This is the essence of the so-called principle of causal closure of the physical world. According to this principle, any physical effect has a sufficient physical cause for its occurrence. The universe contains everything necessary for a complete causal explanation of any of its elements and is complete or closed in the sense that no non-physical causes are needed for such an explanation.

According to this principle, all events in the physical world must have causes that are also physical. This means that every process, every phenomenon we observe, must have its roots in physical laws and conditions. Such a concept ensures the self-sufficiency and closure of physical reality, which does not require the intervention of external or non-physical factors to explain its structure and functioning.

However, despite the fact that the principle of causal closure of the universe is a basic principle of physics, many thinkers have serious doubts that at least some part of reality is causally closed, let alone all of reality.

These doubts are generated by various phenomena and concepts that do not always fit into the framework of purely physical explanations. For example, consciousness is one of the main anomalies we discussed earlier. Its nature and mechanisms of emergence do not lend themselves to simple physical analysis, which calls into question the possibility of a complete causal explanation within the framework of physical closure. Why? Because consciousness exists.

As we have already established, consciousness is an anomaly that does not fit into the framework of physical laws. It cannot be reduced to electrochemical processes in the brain; it is something more. And this "something more" is not subject to the principle of causal closure.

Consciousness is a non-physical cause that has physical consequences. Our thoughts, feelings, desires—all of these affect our behavior, our actions in the physical world. But where do these thoughts, feelings, and desires come from? They cannot be explained solely by physical causes.

If we consider consciousness as a non-physical entity that has the ability to interact with the physical world, then we are faced with a paradox: how can something non-physical have an effect on the physical? This paradox raises serious questions for the principle of causal closure. If consciousness has physical consequences but no physical causes, then this principle cannot be universal and complete.

For example, your decisions and actions can be motivated by emotions, memories, or moral beliefs. Although these phenomena may have certain physical correlates in the brain, their essence cannot be fully explained through physical processes. They arise from subjective experience, which is not subject to physical measurements.

What Am I?

Reflections on this topic can captivate you completely, because your entire personality, everything you know and feel, is contained in the tiny space of your skull. It's like an entire universe existing within the confines of your skull.

Memories: Echoes of the past that shape your present. They can be vivid and detailed, as if you are reliving the events again, or blurry and fragmented, like old photographs. But each memory is a part of you, it's what makes you unique.

Experiences: Accumulated knowledge and skills acquired throughout life. These are your successes and failures, joys and sorrows, lessons you have learned, and conclusions you have drawn. Your experience is your wisdom, it's what helps you navigate the world.

Needs: Basic instincts that ensure your survival and well-being. These are the needs for food, water, sleep, safety, love, recognition. Your needs are your driving force, it's what motivates you to act.

Desires: Dreams and aspirations that fill your life with meaning. These are the desires to achieve success, find love, create a family, change the world. Your desires are your compass, it's what points you in the right direction.

Inner Voice: Your inner dialogue, your thoughts and feelings. It's your advisor and critic, your friend and foe. Your inner voice is your conscience, it's what helps you make the right choices.

All these elements together create your personality, your unique identity. They determine who you are, how you think, how you feel, how you act. But what if all of this is just an illusion created by your brain? What if there is nothing beyond your skull but emptiness?

But what if we dig deeper, beyond our own skulls? In 1970, two German psychiatrists, Ingros and Peterman, decided to conduct an experiment that was supposed to shed light on the nature of human consciousness and its connection to the outside world. They created a "silent chamber"—a room completely isolated from external stimuli: sounds, light, smells. Volunteers who spent a long time in this chamber began to experience hallucinations, loss of sense of time, and other mental disorders.

This experiment, which caused a public outcry, showed how important external stimuli are for our mental health and sense of reality. Our senses are windows to the world through which we receive information about our environment. But what happens when we completely disconnect the brain from all senses? Will anything remain of our consciousness, our personality? Can we even exist without the outside world?

These questions lead us to the second fundamental question:

"What is happening around us?"

This question may seem simple, but in reality, it is much more complex and profound than it seems at first glance. After all, everything we know about the world, we receive through our senses. If we turn them off, will anything remain of our perception of reality? Will we be able to feel, think, exist at all?

The "silent chamber" experiment sparked numerous discussions about its ethics and the possible negative consequences for the psyche of the participants. In 1973, the experiments had to be suspended, leaving many questions unanswered. But this experiment made us think about how much we depend on the outside world and how it shapes our consciousness and our perception of reality.

This question is much broader and more complex than it might seem at first glance. Why is the universe around us the way we observe it? Why do physical laws operate according to these principles? Why is everything the way it is, and not otherwise?

As we have already seen, this entity in the box (your brain) receives information indirectly, through the senses. But you make the assumption that it is at some distance from you, while in reality, it is in your head. Look around: the street, the lamppost, the pharmacy—everything you have ever seen, heard, or felt is just a picture created by your brain that never left its boundaries. An entire universe inside your skull.

Let's consider one of the most interesting paradoxes of modern physics, which further fuels the intrigue around our topic—the Boltzmann brain paradox. This paradox challenges our understanding of entropy and the probability of the emergence of complex structures in the universe.

According to the second law of thermodynamics, entropy—a measure of disorder—in a closed system always increases. This means that over time, the universe should tend towards a state of maximum disorder, where all particles are evenly distributed. But how then can we explain the existence of such complex structures as galaxies, stars, planets, and ultimately, life?

Ludwig Boltzmann, an Austrian physicist of the 19th century, proposed the hypothesis that in a universe with a sufficiently large number of particles and time, random fluctuations can lead to the formation of complex structures, even if they are unlikely (i.e., there is always a probability that anything can form in the universe, even a brain with memories). But according to this hypothesis, a more likely outcome of a random fluctuation is not an entire universe, but only one self-aware brain that spontaneously emerged from chaos and has all the memories and experiences that we have.

This brain, known as the "Boltzmann brain", is a kind of absurdity that challenges our understanding of reality. After all, if the more likely outcome of a random fluctuation is a single lonely brain, then why do we observe a whole universe around us, filled with galaxies, stars, planets, and life?

Of course, this is just a paradox, a kind of thought experiment that makes us think about the nature of reality and our place in it. It emphasizes how little we know about the universe and how many unsolved mysteries remain.

Doubt is the Engine of Progress

Doubt is not only a normal part of human thinking, but also a true engine of progress. It can lead to the discovery of new truths and the unveiling of the unknown. For example, Newton's doubts and his search for truth led to the formulation of the law of universal gravitation, which changed our understanding of the physical world. Similarly, Einstein's doubts stimulated him to create the theory of relativity, which revolutionized our understanding of space and time.

However, doubts can also lead to incorrect conclusions and fantastic theories. For example, the belief in the flat Earth theory or the existence of reptilians shows that doubt can be destructive if it is not supported by scientific facts and evidence.

Nevertheless, it's important to understand that doubt is not just an expression of distrust, but also the key to critical thinking and the search for truth. Doubts can stimulate us to research, analyze, and rethink existing concepts and beliefs. They can lead us to a deeper understanding of the world around us and the development of new ideas and theories.

René Descartes' method, which he formulated in his "Meditations on First Philosophy," played a significant role in the development of philosophy and science in the modern era. This method consists of doubting everything, even one's own existence, and then striving to find an indubitable foundation for certain truth. However, it was in the process of this doubt that Descartes came to the conclusion that the most fundamental and immediate truth for him is that he exists.

Historically, these words "Cogito, ergo sum" ("I think, therefore I am") became a turning point in the development of Western philosophy, as they demonstrate the possibility of using thinking as a way of knowing the world and oneself. Descartes believed that this indubitable truth about existence is fundamental to any understanding of the world.

Consciousness and Science

Consciousness is not just an abstract concept that exists only in philosophical treatises. It is a real, tangible phenomenon that we experience every moment of our lives. And although many believe that consciousness is something ephemeral and beyond scientific analysis, this is far from the truth.

Max Planck, one of the most prominent physicists of the 20th century, considered consciousness to be the fundamental basis of being. He said: "I regard consciousness as fundamental. I regard matter as derivative from consciousness. We cannot get behind consciousness. Everything that we talk about, everything that we regard as existing, postulates consciousness."

How ordinary matter, arranged in a certain way, can give rise to subjective experience is a fundamental question that stumps even the most prominent scientists and philosophers. Roger Penrose, a prominent physicist and mathematician, expresses doubt that natural selection can explain the emergence of algorithms capable of conscious evaluation of other algorithms.

The paradox of consciousness lies in the fact that it is simultaneously the most obvious and the most mysterious thing in the world. We cannot doubt the existence of our own consciousness, but at the same time, we cannot explain it using scientific methods.

Subjective experience is the basis of our knowledge about the world, but at the same time, it remains beyond the scope of scientific explanation. As Schopenhauer wrote, the subject knows everything, but itself remains unknown.

The brain, this three-pound organ, is capable of generating the entire spectrum of human experience, but the mechanism of this process remains one of the greatest mysteries of science.

Imagine that you were disassembled into atoms and then reassembled. Would you remain the same person you were before? Or would it be a different person, with different thoughts, feelings, and memories?

This is a question that has troubled philosophers and scientists for centuries. The problem of personal identity is one of the most complex and mysterious problems in philosophy. It lies in the fact that we, humans, consider ourselves to be the same person throughout our lives, despite the fact that our body and consciousness are constantly changing.

Every day we age, our cells die and are replaced by new ones, our thoughts and feelings change under the influence of experience and circumstances. But we still feel like the same people we were yesterday, a year ago, ten years ago.

What makes us us? What ensures our personal identity? Is it our memories? Our beliefs? Our values? Or maybe it's just our physical embodiment?

There is no single answer to these questions. Philosophers have proposed various theories of personal identity, but none of them are universally accepted.

Some philosophers believe that our personality is determined by our consciousness, our thoughts and feelings. Others believe that our personality is determined by our body, our physical essence. And still others believe that our personality is something more than just consciousness or body, it is something that arises as a result of their interaction.

The problem of personal identity is not just an abstract philosophical question. It has important practical implications. For example, it affects our understanding of responsibility and punishment. If a person has changed, can they be held responsible for their past actions?

The Mystery of Subjective Experience

The mystery of subjective experience remains one of the most intriguing and complex problems in the philosophy of consciousness. The example of the bat, proposed by Thomas Nagel, emphasizes the fundamental impossibility of fully understanding and experiencing the experience of another creature, even if we have detailed knowledge of its biology and behavior. This applies not only to animals but also to other people. Each of us has our own unique inner world, inaccessible to full understanding by others.

The example with the color red demonstrates that subjective experience cannot be reduced to objective physical properties. We can know everything about the wavelength of light, but this will not help us convey the sensation of red to a person who has never seen it. This gap between objective knowledge and subjective experience is a key aspect of the problem of consciousness.

This problem has important implications for our understanding of ourselves and others. It calls into question the possibility of complete mutual understanding and empathy, as we can never fully experience what another person is experiencing. It also emphasizes the uniqueness of each consciousness and the importance of individual experience.

Consciousness - An Epiphenomenon? The Mystery of the Influence of Thought on Reality.

According to epiphenomenalism, our thoughts, feelings, and desires have no real causal power. They are merely epiphenomena, a kind of "exhaust" of brain activity, which do not affect our actions and decisions. Instead, our actions are determined by physical and biochemical processes in the body, which occur independently of our consciousness.

For example, when we experience fear, it is not a conscious decision, but the result of the release of adrenaline and other stress hormones that prepare our body for the "fight or flight" response. Our thoughts and feelings of fear are merely epiphenomena of these physical processes, which have no real impact on our behavior.

Such a view of consciousness undermines our usual understanding of ourselves as conscious agents, capable of making free decisions and influencing the world around us. If our thoughts and feelings have no real causal power, then who is responsible for our actions? Can we even talk about free will if our behavior is entirely determined by physical laws?

Epiphenomenalism raises many questions and doubts. If consciousness has no causal power, then why does it even exist? What role does it play in the evolution and life of a person? Can we even trust our thoughts and feelings if they are just an illusion created by our brain?

Some scientists and philosophers go even further and propose to abandon the concept of consciousness altogether, considering it an unnecessary and misleading concept. They argue that all our mental phenomena can be explained by physical and biological processes, and that consciousness is just a myth that prevents us from understanding the true nature of man.

Daniel Dennett's Theory: Consciousness as an Illusion?

Daniel Dennett, a renowned American philosopher and cognitivist, proposes a radically new view of the nature of consciousness. In his book "Consciousness Explained," he argues that consciousness is not a single, unified phenomenon, but rather consists of a multitude of information streams constantly competing for the brain's attention. These streams, like programs in a computer, determine our behavior, shape our thoughts and feelings.

Dennett compares consciousness to a theater stage where different information streams act as actors, vying for the right to be heard and seen. But unlike traditional theater, where there is a spectator who perceives the performance, in consciousness there is no central "I" that observes this process. Consciousness, according to Dennett, is an illusion created by the brain to unite disparate information streams into a single whole.

Such a view of consciousness has far-reaching consequences. If Dennett is right, then our intuition about the existence of a single "I" is wrong. We are not autonomous subjects who consciously control our lives, but rather a collection of information processes that are constantly changing and interacting with each other.

David Chalmers: The Hard Problem of Consciousness

In 1993, David Chalmers, an Australian philosopher and cognitive scientist, introduced the term "hard problem of consciousness" in his doctoral dissertation, which became a turning point in the philosophy of consciousness. Chalmers argued that even if we could describe in detail the physical and computational processes that occur in the brain, this would not answer the question of why these processes are accompanied by subjective experience. In other words, we can know everything about how the brain works, which neurons are activated during various mental processes, which areas of the brain are responsible for different functions. But this does not explain why we feel what we feel, why we have subjective experience of the world.

The Hard Problem of Consciousness is not just a philosophical puzzle, it's also a major challenge for science. How can we study something as subjective and elusive as consciousness using objective scientific methods? This is a question that neuroscientists, psychologists, and philosophers are grappling with.

Philosophical Zombie: The Possibility of Unconscious Doubles

The concept of the "philosophical zombie," proposed by Eliezer Yudkowsky, challenges the connection between consciousness and physical reality. A zombie is a hypothetical creature that is completely identical to a human on a physical level but lacks consciousness. It can talk, act, and react to external stimuli but has no internal subjective experience.

Yudkowsky argues that if zombies are possible, then consciousness cannot be reduced to physical processes in the brain. It must be something more, something non-physical. After all, if a zombie can exist without consciousness, then consciousness cannot be a necessary condition for the existence of a physical body and its functioning.

This concept has far-reaching implications. If consciousness is a non-physical property, then this opens the door to various metaphysical speculations. For example, it could mean that we live in a simulation created by a more advanced civilization, or that consciousness is a manifestation of some higher reality.

However, Yudkowsky himself considers the zombie argument absurd. He argues that no zombie would write articles about consciousness or argue on this topic because consciousness is required for that. This argument emphasizes the importance of subjective experience for understanding consciousness and questions the possibility of the existence of philosophical zombies.

Problems of Philosophy: The Foundation for Science and Understanding the World

Philosophy is often perceived as an abstract and detached from reality discipline, but this is far from the truth. Philosophy is the foundation for many sciences and fields of knowledge, and its problems have a profound impact on our understanding of the world and ourselves.

Historically, philosophy has been the source of many scientific discoveries and theories. Great thinkers of the past, such as Galileo Galilei, Robert Hooke, and Isaac Newton, considered themselves philosophers. They explored fundamental questions about the nature of space, time, and matter, which at that time were philosophical problems. Their reflections and discoveries laid the foundation for modern physics.

Philosophy deals with complex problems that do not lend themselves to simple scientific explanation. It raises questions about the nature of reality, consciousness, morality, knowledge, truth, and many other fundamental aspects of human existence. Philosophy helps us understand what we know, how we know, and why it matters.

In addition, philosophy plays an important role in defining what is scientific and what is not. It explores the methodology of science, its principles, and limitations. Philosophy helps us understand how science works, its strengths and weaknesses, and how we can use scientific knowledge to solve the problems facing humanity.

Panpsychism: Consciousness as a Fundamental Property of Reality

David Chalmers, a renowned philosopher of consciousness, proposes a radical view of the nature of consciousness called panpsychism. According to this theory, consciousness is not an exclusive property of humans or even animals, but a fundamental feature of reality, present in all things, from elementary particles to complex organisms.

Chalmers argues that traditional materialistic approaches are unable to explain the phenomenon of consciousness because they try to reduce it to physical processes in the brain. But consciousness, in his opinion, cannot be reduced to matter; it is something more, something fundamental.

Panpsychism offers an alternative view of consciousness, considering it an elementary property of the universe that manifests itself to varying degrees in all things. In humans, with their complex nervous system, consciousness manifests itself most vividly, but even elementary particles can have some form of primitive consciousness.

This theory finds support in the words of Erwin Schrödinger, one of the creators of quantum mechanics, who wrote that our consciousness cannot be part of the scientific picture of the world because it is itself this picture.

Panpsychism also resonates with the thoughts of Stephen Hawking, who, in his "A Brief History of Time," noted that science cannot explain why the universe exists and why it has consciousness. Perhaps the answer to this question lies in the very nature of consciousness, which may be a fundamental property of reality.

The Path to Truth

We have embarked on an amazing journey, exploring the boundaries of reality and consciousness. We have considered the simulation hypothesis, pondered the nature of consciousness and its connection to the physical world, and delved into the depths of philosophical problems that have troubled humanity for centuries.

We have seen that consciousness is not just a product of the brain, but something more, something that goes beyond our understanding. It is an anomaly that contradicts the laws of physics and forces us to rethink our place in the universe.

We have pondered the questions "What am I?" and "What is happening around us?" They lead us into the unknown, opening up new horizons of knowledge.

But the search for answers does not end here. On the contrary, it is only beginning. Each new discovery, each new idea brings us closer to understanding the mystery of consciousness and reality.

We can doubt everything, but we cannot doubt our own existence, our own consciousness. This is the only unshakeable truth from which we can start in our search.

Are we living in a simulation? Is consciousness a fundamental property of reality? Will we ever be able to fully unravel the mystery of consciousness?

There are no definitive answers to these questions yet. But it is the search for these answers that makes our lives meaningful and exciting. It encourages us to explore, to reflect, to develop.

Perhaps, to understand consciousness, we first need to understand the origin of everything—the universe, life, and the human itself. Perhaps the answers to our questions lie in the depths of space, in the secrets of evolution, in the genetic code that defines our essence.

In the next chapter, we will delve into the fascinating world of science, exploring the origin of the universe, the emergence of life, and the evolution of humans. We will try to understand how such a complex and amazing thing as consciousness arose from chaos and chance.

So let's continue our journey. Let's explore, doubt, and ask questions. Let's seek the truth, no matter how unusual and unexpected it may be.

CHAPTER 2 : THE UNIQUENESS OF LIFE IN THE UNIVERSE

Earth - The Symmetry Breaker of the Universe

The number of intelligent civilizations in the universe is equal to the number of habitable planets multiplied by the probability of life arising multiplied by the probability of intelligence emerging among life. If you stand up now and look at yourself in the mirror, from the point of view of matter, you will see a significant "imperfection." To understand what this means, I must dispel a very popular myth about matter itself.

Suppose water turns into ice when it freezes, taking on a crystalline form. And all of us from early childhood sincerely believe that ice crystals, snowflakes, or patterns on frozen windows are examples of spontaneously occurring symmetry in nature, examples of the emergence of some kind of harmonious structure in contrast to the prevailing disorder in nature.

But this is not entirely true. In fact, everything is exactly the opposite. A crystal is not the emergence but the breaking of symmetry because water is much more isotropic, that is, much more uniform and symmetrical in all directions, than ice. When water crystallizes, the atoms in it self-organize into distinct patterns, and the space occupied by the crystal loses its symmetry and becomes periodic, with a certain algorithm. This crystallization process disrupts the initial spatial homogeneity of water. Water, which was previously the same in all directions, becomes ice with repeating structures and patterns that are clearly organized in certain directions. Thus, a crystal is not a manifestation of symmetry, but rather its violation, which leads to the appearance of complex and beautiful patterns.

Looking at our entire universe, you will notice the same thing. The universe is quite homogeneous. It doesn't matter where you are—here, there, or anywhere else—on average, you will observe approximately the same picture, regardless of the place of observation. Looking out the window at the starry sky, you will find that the part of the universe we can see is a good example. If you were looking out the window from the other side of the galaxy, you would see roughly the same thing.

But if the universe is so homogeneous, then what are you doing here? Just as ice breaks the symmetry of water, galaxies, stars, and planets break the symmetry of the universe. However, nothing breaks it with its existence as much as a human.

Stanisław Lem, in his monumental work "Summa Technologiae," asks the question: "Is our civilization an ordinary or exceptional phenomenon? Do we correspond to the norms of development accepted in the universe, or are we a deviation, a deformity?"

We understand how and why the process of water crystallization occurs, we understand how and why various cosmic bodies arise, but don't be fooled: we absolutely do not understand how and why we appeared. Our existence, our consciousness, our capacity for self-reflection and creativity—all of this seems like an incredible coincidence, given the vastness and homogeneity of the universe.

The Rare Earth Hypothesis

The Rare Earth hypothesis is based on the idea that the emergence and development of complex life require a combination of many factors that are rarely found simultaneously in space. These factors include not only the presence of water and an atmosphere but also more specific conditions, such as the stability of the star, the optimal distance from it, the presence of a large moon, tectonic activity, and others.

Peter Ward and Donald Brownlee argue that Earth ended up in the "Goldilocks Zone"—a region of space where all the necessary factors came together in a perfect combination. Our planet has a stable orbit around the Sun, which is a star of medium size and age, providing a constant flow of energy. Earth has enough mass to retain an atmosphere, but not so much that it turns into a gas giant. The Moon stabilizes the tilt of the Earth's axis, preventing sharp climate changes. Tectonic activity ensures the circulation of substances and maintains a stable climate.

The authors of the hypothesis emphasize that even slight deviations from these ideal conditions can make a planet unsuitable for complex life. For example, if a star is too active, it can burn the planet's atmosphere or expose it to deadly radiation. If a planet doesn't have a large moon, its axis can oscillate chaotically, leading to extreme climate changes. If there is no plate tectonics, the planet can lose its atmosphere or turn into a scorching desert.

The Principle of Mediocrity vs. The Uniqueness of Earth

How many times have you heard that Earth is just an ordinary rocky planet in a typical planetary system, located in an unremarkable region of one of the countless spiral galaxies? This view, known as the principle of mediocrity or the Copernican principle, is common among many scientists, including Carl Sagan and Frank Drake. They believed that given the vast number of stars and planets in the universe, life must be a common phenomenon.

However, there is an opposing point of view, presented in the book "Rare Earth" by Peter Ward and Donald Brownlee. They argue that Earth is not just an ordinary planet, but the result of an incredible coincidence of circumstances that makes it unique and perhaps the only place in the universe where complex life could have arisen.

The authors of the book emphasize that the Solar System and our galaxy have several features that make them atypical. For example, the Sun is a stable star with moderate activity, providing stable conditions for life on Earth. Jupiter, the largest planet in the solar system, acts as a "cosmic shield," protecting Earth from asteroids and comets. The Milky Way is a spiral galaxy with a moderate number of stars, reducing the risk of supernova explosions and other cosmic catastrophes.

A planetary system capable of supporting complex life must indeed have a planet in the so-called "Goldilocks Zone"—the region around a star where the temperature allows for the existence of liquid water. This condition is necessary because water is a universal solvent and the medium for many biochemical reactions that underlie life as we know it.

The uniqueness of Earth lies not only in its location in the Goldilocks Zone but also in the stability of its orbit. Thanks to its nearly circular orbit, Earth never moves so far away from the Sun that water freezes, and it never gets so close that it evaporates. This ensures a stable climate necessary for the development and maintenance of life.

However, while having a planet in the Goldilocks Zone is a necessary condition for life, it is not sufficient. There are many other factors that also play an important role. For example, the composition of the atmosphere, the presence of a magnetic field, tectonic activity, the influence of moons, and others.

Even if a planet is in the Goldilocks Zone but does not have a protective magnetic field, it will be bombarded by cosmic radiation, which can hinder the development of life. If a planet does not have tectonic activity, it will not be able to maintain a stable climate and the circulation of substances necessary for life.

The Galactic Habitable Zone

But for life, as you will soon understand, this is catastrophically insufficient. From the point of view of the authors of "Rare Earth," the habitable zone within its star is not the only habitable zone in which it needs to be. Our entire solar system was also lucky to be in the habitable zone of the entire galaxy.

Our solar system revolves around the center of the galaxy with a speed of one full revolution in approximately 235 million years. This means that since the appearance of oceans on Earth, it has traveled this entire distance 17 times, since the appearance of the first reliable traces of life - 15 times, and since the appearance of the first multicellular organisms - about 7 times. And this is not just a relaxing flight, you know. The vastness of the galaxy holds many terrible things, absolutely deadly for our tiny solar system, especially over such a vast period. However, life is still here. Do you understand?

Dead Zones of the Galaxy

Most of any galaxy is a dead zone, unable to support complex life. The closer to the center of the galaxy, the stronger the harmful effects of X-ray and gamma radiation generated by black holes and neutron stars, which are more numerous the closer we are to the galactic center. The closer we are to the galactic center, the more often supernova explosions occur nearby—another pleasant thing that can destroy entire worlds at a distance of many light-years.

There are a lot of stars in galaxies, and even more planets around these stars. This means that life should be teeming throughout the Milky Way. But when we say that, we are most likely mistaken because the bulk of the stars are concentrated closer to the middle (which is not surprising), and that's where the real meat grinder happens.

In addition, the high density of stars closer to the center or in the spiral arms of the galaxy means large gravitational perturbations. That is, even if a planet orbits its star at an optimal distance, all this fine-tuning will be disrupted when another star or something worse flies nearby. In general, being too close to the center of the galaxy is bad.

However, being too far away is also bad. With distance from the galactic center, the metallicity of stars decreases, and metals are extremely necessary for the formation of terrestrial planets.

Recent studies suggest that the extent of the galactic habitable zone is about 7-9 kiloparsecs from the center of the galaxy. But to put it simply, no more than 10% of the stars in the Milky Way fall into this zone. Other studies suggest that only 5% of stars are suitable for the origin of life.

Therefore, a planetary system suitable for life must stably maintain favorable conditions for billions of years for complex life to have time to develop. Therefore, it is desirable for a life-bearing star to have an almost perfectly circular orbit around the center of the galaxy, and in addition, its rotation speed should be synchronized with the rotation speed of the spiral arms, so that, as you understand, it passes through dense clusters of stars as rarely as possible. And our solar system is precisely located in the habitable zone of the entire galaxy. The Sun's orbit around the center of the Milky Way is almost perfectly circular, with a rotation period of 235 million years, which roughly corresponds to the rotation period of the galaxy.

Our planet was lucky to be in the Goldilocks Zone of its star system, which, in turn, was lucky to be in the Goldilocks Zone of its galaxy. But that's not all.

The Uniqueness of the Milky Way

Our cosmic home, the Milky Way, turns out to be not just another spiral galaxy among billions of others. It is a true exception, a unique formation that played a key role in the emergence of life on Earth.

Consider this: for ten billion years, the Milky Way has not collided with other galaxies. This is a phenomenal period of cosmic tranquility that has allowed our galaxy to avoid catastrophic collisions that could have led to the destruction of planetary systems and the extinction of potential life.

Such prolonged isolation has made the Milky Way anomalously quiet and dim. It lacks active galactic nuclei that emit powerful streams of energy, and other sources of cosmic radiation that could be harmful to life. Only seven percent of galaxies in the universe can boast such a peaceful history.

This unique feature of the Milky Way has a direct impact on the possibility of life's existence. After all, high levels of radiation and frequent cosmic cataclysms can hinder the formation of stable planetary systems and the development of life on them.

What's Wrong with Half the Stars in the Cosmos?

The familiar image of a single star, like our Sun, turns out to be not as common as we used to think. More than half of the star systems in our galaxy are binary or multiple, meaning they consist of two, three, or even more stars orbiting a common center of mass.

This discovery challenges the notion that planetary systems like ours are typical in the universe. In binary and multiple systems, the gravitational interaction between stars creates complex and unstable conditions for the formation and existence of planets.

Planets in such systems can have chaotic orbits that constantly change under the influence of the gravity of several stars. This can lead to extreme temperature fluctuations, collisions with other celestial bodies, and other catastrophic events that make life on such planets practically impossible.

Even if a planet in a binary or multiple system is in the habitable zone of one of the stars, its orbit may be unstable and change over time, leading to the planet leaving this zone. In addition, the gravitational influence of other stars can cause tidal forces that lead to volcanic activity, earthquakes, and other disasters.

The Uniqueness of the Sun

Our Sun, although it seems like an ordinary star in the night sky, actually has a number of unique characteristics that make it an ideal source of energy for life on Earth.

First, the Sun is an extremely stable star. Its luminosity changes by only one-tenth of a percent, ensuring a constant and uniform flow of energy to Earth. This is critical for maintaining a stable climate and the conditions necessary for life.

Second, the Sun has an optimal mass. Stars with a larger mass than the Sun live much shorter lives, quickly burning out and turning into supernovae. Stars with a smaller mass, red dwarfs, although they live longer, emit little energy and have narrow habitable zones, making the conditions on planets around them unsuitable for life. The Sun, with its mass, provides enough energy to support life for billions of years.

Third, the Sun is a yellow dwarf—a type of star that has the optimal temperature and size for creating a wide habitable zone. Planets in this zone can have stable orbits and a climate favorable for the development of life.

Fourth, to date, no twin star of the Sun has been found in terms of size, luminosity, temperature, age, and metallicity. The metallicity of a star is the content of elements heavier than helium in it. High metallicity is necessary for the formation of planets, as well as for ensuring the stability of the star's luminosity.

Red Giants vs. Red Dwarfs

The larger the star, the wider its habitable zone, and therefore, the higher the probability of the emergence of life itself. Well, yes, if we are talking about temperature. However, the larger the star, the more powerful the level of ultraviolet radiation, and the faster it will burn all its fuel and turn into a red giant. For example, our beloved Betelgeuse is about to die, having lived only about 10 million years. For comparison, the Sun is 500 times more long-lived.

Disregarding radiation, even the simplest life near supermassive stars most likely simply won't have time to develop.

Let's then take red dwarfs, the most common type of star in the universe. They seem to be attractive candidates for the search for life. They have weak radiation, which reduces the risk of harmful effects on biological molecules, and an incredibly long life cycle, reaching trillions of years. Unlike massive stars, red dwarfs never turn into red giants, making them more stable neighbors for planets.

However, their low temperature and dimness create significant obstacles to the development of life. To receive enough heat and light, a planet must be very close to a red dwarf, which leads to two main problems:

- **Tidal Locking:** At such a close distance, the planet falls into the star's tidal lock, forcing it to rotate synchronously with its orbital motion. This means that one side of the planet always faces the star, and the other is always turned away, leading to extreme temperature contrasts. Even if the temperature in the terminator zone (the strip between light and darkness) could theoretically be suitable for life, the strong stellar wind of the red dwarf, which is not much weaker than the solar wind, destroys any chances of its development.
- **Super-Earths:** Observations of exoplanets around red dwarfs have shown that most of them are super-Earths—planets that are significantly larger than Earth but smaller than gas giants. Such planets have much stronger gravity than Earth, which can complicate the development of complex life forms. In addition, super-Earths often have a dense atmosphere, which can create a greenhouse effect and prevent the formation of a stable climate.

These problems make red dwarfs less attractive candidates for the search for life than Sun-like stars. Although they have some advantages, such as longevity and stability, their disadvantages make them less favorable for the development of complex life.

The Uniqueness of the Planets in the Solar System

Our solar system, which we often consider typical, actually has several features that make it quite unusual compared to other known planetary systems.

First, the solar system is distinguished by a huge variety of planets. From tiny Mercury to the gas giants Jupiter and Saturn, our planets have different sizes, compositions, and orbital characteristics. This contrasts with most known exoplanetary systems, where super-Earths—planets that are in size between Earth and Neptune—predominate.

Second, the presence of two gas giants, Jupiter and Saturn, with relatively circular orbits at a great distance from the Sun, is also a rare phenomenon. Gas giants are found in only 10% of known planetary systems, and they usually have more eccentric orbits and are located closer to their stars.

The origin of such an unusual architecture of the solar system still remains a subject of scientific debate. There are various theories that try to explain how our planets formed and why they have such different characteristics.

One theory suggests that the gas giants initially formed closer to the Sun and then migrated to their current orbits, scattering or absorbing smaller planets that were in their path. Another theory argues that the gas giants formed in their current orbits, but their formation was a unique process that is not repeated in other planetary systems.

Unsolved Mysteries of the Solar System

And how many variables remain unclear? After all, the authors of "Rare Earth" list only what is more or less on the surface. For example, the huge Earth's core. For some reason, it is larger than expected, and probably thanks to it, we have a magnetic field that protects us from the solar wind and cosmic rays. If there were no magnetic field, the Earth's atmosphere would be blown away by the solar wind, as happened with Mars.

Or take the Moon. It is not just a satellite; it stabilizes the Earth's axis of rotation, ensuring a stable climate. Without the Moon, the Earth could tumble like a top, leading to catastrophic climate changes.

Or here's another one: Earth is at the perfect distance from the Sun so that the temperature on its surface is high enough for the existence of liquid water, but not so high that the water evaporates.

And these are just some of the factors that make our planet unique. Perhaps there are many other unknown factors that also played a role in the emergence of life on Earth.

Just look at the other rocky planets in the solar system. Mercury and Venus have no moons at all. Look at the moons of Mars, Phobos and Deimos. They are not even spherical because they are not moons, but simply asteroids captured by gravity. The diameter of Phobos is 22 kilometers, and Deimos is only 12.

On the other hand, Earth's Moon is the largest natural satellite in the solar system relative to the size of its planet. It is 27% of the size of Earth. Pluto, which until 2006 was considered a classical planet, is about six times less massive than the Moon.

The formation of such a satellite is a mystery in itself. It was discovered that samples of lunar rock taken by Apollo astronauts are suspiciously similar in composition to the Earth's crust. Therefore, the best theory we have so far for the Moon's appearance is the giant impact hypothesis. This is a hypothetical ancient planet the size of Mars, which allegedly collided with Earth about four and a half billion years ago. Upon impact, the core and mantle of Theia mixed with the core and mantle of Earth, so it is quite likely that we have two planets here under our feet. Hence the large Earth's core.

The Influence of the Moon on Earth

And what about the debris from the collision that formed the Moon? So what? The fact that such a large satellite provokes tides. Darwin himself suggested that the simplest life first originated in one of the basins that arose as a result of the tidal interactions of the Moon. In addition, some scientists believe that after the emergence of life in water, its further spread from the ocean to land was also provoked by the gravitational influence of the Moon.

Four billion years ago, the Moon looked very different in the sky than it does now. Its gravitational influence was much stronger, causing completely different tides than those we are used to. Giant tides threw living organisms onto land, where they died. This process continued until some organisms developed beneficial mutations that allowed them to survive in these conditions. This led to the gradual adaptive transition of life from the aquatic environment to land.

But even the oceans themselves on the planet... Data obtained in 2019 suggests that Theia could have formed in the outer solar system, and much of the Earth's water originated on it. So, perhaps, the oceans you see are not native to these places.

The authors of "Rare Earth" note that a large satellite increases plate tectonics, which is necessary for biodiversity and the carbon cycle. This is not to mention that the Moon, like a dynamo, contributes to the Earth's strong magnetic field, which protects us from cosmic radiation.

In addition, this giant impact gave the Earth a special axial tilt and a fast rotation speed. The rapid rotation reduces daily temperature fluctuations and makes photosynthesis viable. According to the Rare Earth hypothesis, the axial tilt should not be too large or too small, because a planet with a large tilt will experience extreme seasonal climate fluctuations. If the planet has a small tilt or none at all, there will be no seasons on the planet, which will lead to (as funny as it sounds) a lack of incentives for evolution. From this point of view, the Earth's tilt is just right. The gravity of a large satellite also stabilizes the planet's tilt. Without this effect, the change in tilt would be chaotic, which would likely make the existence of complex life forms on land impossible.

For example, throughout Earth's history, there have been five major mass extinctions that wiped out 75% to 96% of all species. And although these events were catastrophic, they also played an important role in the evolution of life. They freed up ecological niches, allowing new species to evolve and thrive. If there were no mass extinctions, perhaps life on Earth would have remained at the level of simple single-celled organisms.

Another paradox is related to the size of the Earth. As we have already said, Earth has the optimal size for supporting life. But what if it were a little bigger or a little smaller? It would seem that a small change in size should not matter much. But in reality, even minor changes in the size of a planet can have dramatic consequences for its climate and ability to support life. For example, if Earth were a little larger, its gravity would be stronger, leading to a denser atmosphere and a greenhouse effect. If Earth were a little smaller, its gravity would be weaker, and it would not be able to retain the atmosphere necessary for life.

Another important factor is the composition of the Earth's crust. Earth has a unique crustal composition, rich in elements such as silicon, oxygen, aluminum, iron, calcium, sodium, potassium, and magnesium. These elements are necessary for the formation of rocks, soil, water, and other components necessary for life. If the composition of the Earth's crust were different, perhaps life on Earth could not have arisen.

Finally, we must not forget the role of chance in the evolution of life. Many events that led to the emergence of humans were the result of random mutations and coincidences. If it were not for these coincidences, perhaps life on Earth would have taken a completely different path.

Paradoxes of Evolution

Throughout Earth's history, the temperature has dropped drastically twice, and it was completely covered in ice: 2.2 billion and 635 million years ago. And what do you think? Such extreme glaciations contributed to the development of life! The first stimulated the development of photosynthetic microorganisms, which led to a sharp decrease in the level of greenhouse gases in the atmosphere and the release of oxygen. And after the second glaciation, the Cambrian explosion occurred, thanks to which almost all evolutionary branches of animals existing today arose. If before this all life was simple and unicellular, then after, there was a sharp increase in the number of complex multicellular organisms. And the brain appears right here.

Think about it: if we find a planet in space with paradise conditions, it is likely that we will not find complex life there precisely because of that. It is simply not needed on a paradise planet. There are no complex survival challenges that need to be solved. As we can see, the development of complex life requires contrast and challenge.

Knowing only the positive factors is not enough. They must be balanced by negative ones for the same stimulation of evolutionary processes. But at the same time, negative factors should not be too negative, because life is a rather fragile thing.

How many factors in total have to come together in the right sequence for complex life to arise, and ultimately, intelligence? This question remains open.

The Uniqueness of the First Replicator

Let's start small. Suppose we have all the necessary conditions for the existence of life: the Goldilocks Zone within a star and galaxy, gas giant protectors, a rocky planet with water, the right atmosphere, and a large moon. Imagine that somewhere in space there is an absolute twin of Earth. What's next? Nothing, if we don't consider the fundamental component of life—the replicator.

The first, smallest "building block" of life is a replicator, that is, a molecule capable of self-reproduction. Only with the appearance of the first replicator do the directed mechanisms of evolution begin to operate. But this first replicator itself had to appear by chance. Think about it: a complex system capable of copying itself had to arise by chance. Therefore, we are looking for the simplest combinations from which this first replicator could have emerged. However, even the simplest combinations we see are still too complex for blind chance.

So, what is the probability of the appearance of the first simplest organism capable of reproduction? We are talking about an organism so simple that if you add one more element, it will no longer be able to reproduce. A well-known expert in evolutionary biology, Eugene Koonin, in his book "The Logic of Chance," gives an estimate of the probability of the spontaneous appearance of the first replicator. He calls the figure 1 to 10 to the power of 1018 (that's one with one thousand and eighteen zeros).

It is practically impossible to comprehend how unlikely this event is. Eugene Koonin himself sees one of the possible explanations for this incredible probability in the existence of the multiverse. If there are an infinite number of universes, then there will be an infinite number of configurations of atoms. In this case, the realization of the option with the spontaneous appearance of a replicator becomes more understandable.

However, it would be very convenient to immediately explain such things by the multiverse or the world will or God, and not rack our brains anymore. Therefore, Koonin reminds us that we should lower the bar as much as possible and explain by spontaneous emergence only what really cannot be explained otherwise. The emergence of life requires extremely unlikely events, and therefore, we may be the only living beings in our universe.

We are not only talking about intelligent beings, but about living organisms in general. Of course, we should not exaggerate. But understanding the uniqueness of the first replicator helps us appreciate how important and rare life is as we know it.

Evolution is Not Enough

So, we have the first replicator, which begins to multiply. Here a very important question arises: is the growth of complexity in evolution mandatory? The answer is not at all mandatory. Some of the first replicators—prokaryotes (bacteria and archaea)—were extremely inefficient in terms of energy due to their design. They could not accumulate energy in large quantities. In order to become more complex, it was impossible to get by with just some mutation. This is the moment when evolution was not enough—a revolution was needed.

Because of this, at the level of prokaryotes, development stopped for a billion years, and maybe even more. Nothing foreshadowed the emergence of more complex life forms. Evolution could simply stop at this stage. But suddenly something happened—something as unlikely as the appearance of the first replicator. Something that, according to British biochemist Nick Lane, could only happen once in the entire history of life. An archaeal cell engulfed a bacterium, and for some reason, did not digest it.

This bacterium, in the process, lost 99% of its genome, got rid of most of its mechanisms, and focused on producing energy in the form of ATP. Thus, the first mitochondrion appeared. The mitochondrion is a kind of power plant of the cell. It was this symbiosis, having solved the energy problem, that made it possible for life to become more complex unhindered. Thanks to this microscopic incredible event, all the diversity of life that you can see with the naked eye exists.

The Number of Intelligent Civilizations

I hope no one is tired yet, because the most interesting part is still ahead. Now we come to the question of the number of intelligent civilizations in the universe. We have already discussed how unlikely the key events in the history of life on Earth were. But even if there are other planets in the universe where conditions for life have arisen, this does not mean that intelligent beings have necessarily developed there.

In search of answers to these questions, scientists turn to the famous Drake equation, proposed by astrophysicist Frank Drake in 1961. This equation attempts to estimate the number of intelligent civilizations in our galaxy capable of communication. It includes many factors, each of which has significant uncertainty. Among these factors are:

- The average rate of star formation in our galaxy. This is the basic factor that determines the number of new stars appearing over a certain period.
- The fraction of stars that have planetary systems. Recent observations show that many stars have planets, which adds optimism about the potential for life.
- The number of planets per star system where conditions suitable for life may exist. This question includes the so-called "Goldilocks Zone," where the temperature is suitable for the existence of liquid water.
- The probability of the emergence of life on such a planet. This factor remains one of the biggest mysteries, as we still do not know exactly how life originated on Earth.
- The probability of evolution from simple to complex organisms. The evolution of complex life forms requires billions of years and a number of favorable conditions.
- The probability of developing a technologically advanced civilization. This factor determines whether intelligent life will develop to a level capable of interstellar communication.
- The average lifespan of such civilizations. This is a key point, as civilizations can self-destruct or experience disasters that limit their ability to exist and communicate for long periods.

Each of these factors has its own probability, and together they create a very complex equation with a high level of uncertainty. Therefore, even if there are a large number of planets suitable for life in the universe, this does not guarantee that intelligent beings have developed on them.

Given all these unlikely events, it can be assumed that intelligent life may be a very rare phenomenon. Perhaps we are the only intelligent beings in our galaxy, or even in the entire visible universe. This realization adds even more weight to our every step, every discovery, and every attempt to understand our place in this vast and complex cosmos.

The Uniqueness of the Mind

The mind is not just a consequence of random evolutionary processes. Even if we have many planets suitable for life, even if life arises on these planets, this does not guarantee the emergence of the mind. If all the previous reasoning was mostly theoretical, then we can observe this factor with our own eyes here and now.

Humanity has described over 1.7 million species of living organisms. Of course, this is not all: the total number of species existing on Earth at the moment, according to some estimates, reaches 8.7 million. And throughout the history of the planet, there may be about 500 million of them. Is this a lot or a little? And how many of them are intelligent?

The conclusion of the previous section was that we hardly know what mind, intellect, and consciousness are. We only have some intuitive understanding. But, based on our experience, we can conclude that the emergence of the mind is possible. However, on the example of all other life on the planet, we see that evolution is not very eager for the mind.

Researcher Charles Lineweaver argues that when considering any extreme traits in animals, intermediate stages do not necessarily lead to inevitable results. A large brain is no more inevitable than long noses in animals such as aardvarks and elephants. Monkeys, whales, elephants, dolphins, octopuses, and squid are considered the most intelligent creatures on Earth after humans. Dolphins, in particular, are very intelligent, their brains are in many ways similar to human brains. But, as Lineweaver notes, dolphins have had about 20 million years to build a radio telescope, but they haven't. Why?

Why is the only so-called intelligent species on the planet us? Doesn't it surprise you that evolution, which always strives for economy in its manifestations, created the human brain—a device with such a degree of redundancy that it now, in the 21st century, copes well with the problems of a developed civilization? After all, anatomically and biologically, it is the same brain as the brain of our primitive ancestor, who lived 100 thousand years ago. How could this enormous potential of the mind, this redundancy that was ready at the dawn of history to begin building civilization, arise in the course of a purely probabilistic evolutionary game with the addition of two vectors: an increase in the number of mutations and an increase in natural selection?

The Probability of Intelligence

What is the probability of each variable in this equation?

As Eugene Koonin noted, for the emergence of the first replicator, the probability was extremely small, about 1 to 10 to the power of 1018. However, as our experience shows, the mind is an even more unlikely phenomenon. A large brain capable of self-awareness, abstract thinking, and understanding its place in the universe is not a mandatory product of evolution. It is an exceptional, perhaps unique phenomenon.

The famous astrophysicist Frank Drake, having created his famous equation, included the factor of intelligence as one of the least defined parameters. Even if life arises frequently in the universe, intelligent life may be extremely rare. This leads us to the conclusion that we may be the only intelligent beings in the entire galaxy or even in the visible universe.

Therefore, our existence as an intelligent species is the result of a series of unlikely events. We are the result not just of evolution, but of evolution reinforced by exceptional conditions and circumstances. This makes our minds even more unique and valuable. And perhaps our task as an intelligent species is to use this unique gift to understand not only the world around us but also ourselves and our place in this vast and mysterious universe.

The Probability of an Earth-Like Planet

The probability of the existence of intelligent civilizations in space depends on many factors that we have tried to list. And what do we have? To some extent, we understand how many factors are needed and what they should be for a planet to become suitable for life. At the same time, we have a very weak idea of how many factors must converge for the simplest form of life to appear on a planet suitable for life. And we do not understand at all what factors, in what quantity, and in what sequence are needed for the emergence of intelligence on a planet suitable for life from the simplest form of life.

The absolute majority of factors, as I have already described, are unknown to us. But let's dream. Suppose we have a formula that guarantees a 100% emergence of complex life if only, say, a meager 80 factors are met. Very optimistic. But let's make it even more optimistic: let the probability of each factor be not one in ten to the 1018th power, but only one in two, as when tossing an abstract coin with zero thickness. That is, for example, the probability of finding a sun-like star is not one in a thousand or a million, but fifty-fifty. The probability of finding a rocky planet near such a star is fifty-fifty. Fifty-fifty is the probability that it will be in the habitable zone, that it has a moon-like satellite, a suitable atmosphere, and even the probability of the emergence of the first replicator on it is the same 50/50. And so on, and so forth, until we collect all 80 necessary parameters.

If everything were like this in the universe we see, what would be the probability of meeting brothers in mind in such a scenario? The probability is 10 to the minus 24th power, or one in a trillion trillion. In simple words, the probability is so extremely small that you can search the entire observable universe but never find a single planet like Earth, not one.

In reality, there are likely many more factors, and it is unlikely that even one of them has such an optimistic probability as when tossing a coin. We often hear that it is too selfish to claim that we are alone. But this is a pompous opinion based on general assumptions. We do not yet have any confirmation of the existence of extraterrestrial life, not only intelligent but any at all.

What's Next?

Although we still do not have definitive answers to the question of the existence of other intelligent life in the universe, it is important to continue research in this direction. Science is constantly expanding our knowledge of the cosmos, and each new discovery brings us closer to solving this great mystery. Regardless of whether we are the only intelligent beings in the universe or not, the very process of searching and reflecting on this question expands our horizons and gives us a new understanding of our place in the vast cosmic context.

However, even if we accept the Rare Earth hypothesis as a working model, this does not mean that we should stop there. On the contrary, it opens up new questions and new directions for research. If life on Earth is truly a unique phenomenon, then what factors contributed to the emergence of consciousness and intelligence?

Although we have more or less figured out how life originated on Earth, it is now necessary to talk in more detail about how it evolved to understand our consciousness. After all, it was evolution, this long and complex process, that shaped our consciousness, our mind, our ability for self-awareness and creativity. Understanding the evolution of life can be the key to unraveling the mystery of consciousness and help us answer one question: "What am I?"

CHAPTER 3 : THE EVOLUTION OF LIFE AND CONSCIOUSNESS

The Paradox of Life and Consciousness

Life. Do you know what life is? It's an amazing phenomenon that arose amidst the cosmic void. It's a fragile spark that flared up for a moment in the boundless expanse of space and time. It's a unique creation of nature that, despite its fragility, is able to resist the forces of entropy and chaos.

The most mysterious phenomena in the universe—black holes and dark matter—are not the most mysterious. We see traces of their existence everywhere. Against the backdrop of the entire cosmos, it is life that looks like some kind of anomaly. We have not encountered manifestations of life anywhere outside our planet. Among hundreds of billions of kilometers of emptiness, rock, and gas, something so strange appears that it can oppose everything else. We, living people, are flesh of the flesh of the universe that we see, and we are certainly a part of it. At the same time, the universe is inanimate. It seems so to us. Have you ever thought about it? What is the fundamental difference? What are the differences between living and non-living matter? Where is the line beyond which one ends and the other begins?

If you look through the eyes of an ancient person, the difference is obvious: inanimate matter is motionless, passive, subject to external forces. A huge stone in the dark may well resemble a living bear. But only living matter, a living bear, can chase you through the forest. And no matter how you rush among the trees, catch you and tear you to pieces—inanimate objects never do anything so purposeful. Stones rolling down a mountain, a huge boulder, are not really chasing you. This is how the ancients reasoned. No wonder, because they had never seen homing missiles.

And yet, despite the fact that most of us spend most of our lives doing things we don't like, we are still alive. Alive because we do it of our own free will. Because we have a choice. Because we can say "no."

A stone falling from a mountain has no choice. It obeys the laws of gravity. A uranium atom decays, obeying the laws of quantum mechanics. And we, living beings, can choose. We can go against the current. We can create our own laws.

This is what makes us alive. Not some magical spark, not a special substance, but the ability to choose, the ability to act according to our desires and goals. This is the very foundation of the living, the first step from which everything begins.

Of course, our choices are not always free. We are influenced by genes, upbringing, and environment. But even within the strictest framework, we can always find room for maneuver. We can choose how to react to events, how to interpret information, how to build our relationships with other people.

The Structure of Life

Imagine a pile of stones that are thrown into the air over and over again. There is a chance that the stones will fall on top of each other, forming a certain structure. And this structure will have properties that a simple pile of stones does not have. It will be something more than just the sum of its parts.

This phenomenon is called emergence. Emergence is the appearance in a system of properties that are not inherent in its elements separately. There are many examples of such synergy in nature. By connecting certain atoms in a certain sequence, we get a structure with properties that are not present in a simple pile of the same atoms. This structure is a molecule.

The basis of life is molecular, and the distinctive feature of living molecules is that they provoke the environment to copy them. Some configurations of atoms are prone to replication, that is, to creating copies of themselves using the environment. We call such configurations genes. A gene is a property of atoms, the same property as the complex patterns of a snowflake or ripples on a sand dune.

Piece by piece, gathering together, atoms combine into molecules, which, in turn, form whole organisms. The same "magic" that people tried to find in living organisms is nothing more than a manifestation of the magnetic properties of matter, when elementary processes at the micro level collectively create incredibly complex multi-level phenomena. For the same reason, people who are looking for a center of consciousness in the brain have a very poor understanding of what they are doing at all.

So, once on planet Earth—no one still knows exactly how and where—matter exhibited an extremely rare and very unlikely elegant property. Namely, under uniquely favorable conditions, the first set that began to multiply spontaneously appeared from a pile of atoms. And that's it. Having uncontrolled reproduction with limited resources, we get competition.

Self-copying does not happen perfectly, due to which mutations occur. Mutations give rise to variations of genes, which, in turn, give rise to variations of species. Competition plus variations of species, plus natural selection, plus time equals evolution. Yes, all life on the planet is based on replica molecules that we call genes. Thus, in the universe, there is no division into living and non-living matter. Matter is the same, and life is simply its elegant property.

But does this answer the question of what life is? Don't rush. A gene is the smallest and indivisible element of hereditary material that can be passed from parents to offspring. Genes determine the entire organism: what it will be, what its functions, features, and properties will be. Genes are the instructions by which the entire organism is built.

Richard Dawkins, in 1976, published the book "The Selfish Gene," in which he greatly deepened the theory of evolution. Unlike Darwin, who believed that the most adapted variant of the species survives, Dawkins argues that in fact, the most adapted variant of the gene survives. Thus, all evolution ensures the well-being of the genes that are copied best. So, the unit of evolution is the gene, not the organism, as many believe. For example, the genes of a huge peacock, its tail, clearly do not make life easier for an individual peacock or the entire species as a whole. But it is these genes that have proven to be the most successful compared to their direct competitors. Hence the name—"selfish gene."

The Gene as the Driving Force of Life

The main feature of a gene is that it is always the cause of its copying, that is, it forces the environment to copy it. Thus, biology is, think about it, the science of studying the influence of genes on other matter.

What is a bear? It is a huge container for genes. Take, for example, the insulin gene. It cannot replicate itself. The presence of all other genes in the bear's body is critically important to it. But that's not all. It also needs the genes of other organisms. In the external environment, a bear cannot survive without food, and the genes for producing this food are present only in other organisms.

Therefore, genes jointly construct the body of a bear, prompting it to do many different things, including hunting, so that they, the genes, can copy themselves. The bear's body is the closest environment that genes manipulate for their own replication. They create the bear and program its behavior, forcing the animal to follow algorithms aimed at copying. No, not the bear, but the genes that are in its body. And to do it at any cost.

Note again, the bear itself is not the goal. During reproduction, the bear is not copied, and even less does it prompt its own copying. A new bear is created anew according to the blueprints embedded in the DNA of the parent organisms. The organism itself is not the author; it is part of the replication environment, usually the most important part after all the other genes. The rest of the environment is the type of habitat that the organism can occupy, such as the top of a mountain or the bottom of the ocean, and the specific way of life in it. For example, a hunter or gatherer, which allows the organism to live there long enough for its genes to replicate.

Thus, the organism is the closest environment manipulated by the real authors, that is, the genes of this organism. They force the organism to act in their interests, ensuring their own survival and reproduction.

The Organism as an Environment for Gene Replication

Previously, for example, a bear's nose and its den would be classified as living and non-living objects, respectively. But there is no fundamental difference at the basis of this division. The role of a bear's nose is not fundamentally different from the role of its den. Neither is self-sufficient, although new examples of both are constantly being created. Both the nose and the den are just parts of the environment that are manipulated by the bear's genes in the process of their replication.

Genes that require cooperation for replication combine into long chains known as DNA. Based on these chains, various organisms are formed, such as microbes, plants, or animals. We are used to thinking of these organisms as alive. But from all that has been said, it follows that the term "living" in relation to parts of the organism other than DNA is, to put it mildly, an exaggeration.

To put it simply, it turns out that the main part of living organisms is DNA, which contains genes. All other parts of the organism—cells, tissues, organs—are just tools that help this DNA perform its function, that is, to survive and reproduce. The organism as a whole is a complex machine built by genes to ensure their survival. Therefore, when we talk about life, we actually mean first of all DNA and genes that carry information. All other parts of the organism are auxiliary mechanisms created to support this DNA in its environment.

This understanding of life, in which organisms are viewed in relation to genes as part of the environment, was contained in the foundations of biology since Darwin's time, but it was almost unnoticed until the publication of Richard Dawkins' works "The Selfish Gene" (1976) and "The Extended Phenotype" (1982).

The Gene as a Quantum of Information

Ask yourself the question: what is the basis of life? The answer to this question is the gene. But what is a gene really? A gene is a piece of hereditary information, a unit that carries instructions for the formation and functioning of living organisms. A bear, a human, flowers, bacteria—everything we call life—are just different ways in which genes copy themselves. Ways. But why do some genes survive and thrive while others disappear in the flow of time?

The more accurate information about the environment a gene contains, the more successful it is in terms of its reproduction. Because, having reliable information about the environment, a gene can live and reproduce more effectively in this environment. That is, genes with more accurate information about the world displace genes with less accurate information. This is the fundamental principle of evolution: it is not the strongest or the smartest that survives, but the one who is best adapted to their environment.

The absolute majority of genes containing incorrect or inaccurate information about the environment could not survive in it long enough and died out. The most adapted to the environment survives. But what is adaptation to the environment? Adaptation to the environment is, in essence, the presence of reliable knowledge, that is, objective information about this very environment. It's not that genes can "know" anything in the traditional sense, but evolution works through selection: by trial and error, those genes that have less complete and less reliable information are eliminated.

Imagine evolution as a huge computer algorithm that continuously tests different combinations of genes. In this process, only those that provide a better match to environmental conditions are preserved. It's like a game with billions of variables, where success depends on how accurately genetic information reflects reality.

This selection process leads to the fact that modern living organisms carry genes that best reflect the environment in which they evolved. This allows organisms to find food more efficiently, avoid predators, reproduce, and survive. The accuracy of this information directly affects the chances of survival. Thus, we can say that life on our planet is a continuous process of improving the "information quanta" contained in genes.

The Brain as the Most Perfect Information Processing Tool

Genes exercise very precise interactive control over the organism's responses to the complex environment with which it interacts. This control is aimed at causing a very specific corresponding action of the environment on the genes, namely, to replicate them. Moreover, the more situations outside the genes are able to "digest" without dying, the more successful they are. In this context, the human brain is potentially the best that genes could build. From what we know, the human brain is the most complex and efficient information processing tool in the world of living organisms.

The human brain consists of billions of neurons, each of which is capable of forming numerous connections with other neurons, creating an incredibly complex and dynamic network. This network allows the brain to process huge amounts of information, learn, make decisions, and adapt to changes in the environment. Thanks to its ability for abstract thinking, creativity, and awareness, the brain can model future events, plan actions, and even create new technologies and ideas that change the very nature of our existence.

The more information in genes corresponds to the conditions of the environment, the more copies of these genes will remain. Thus, we can say that evolution is, in essence, a certain process of computation. It is a process in which the most effective survival and reproduction strategies are selected based on the information embedded in the genes. Genes are literally embodied information about the environment, which is constantly being improved and adapted through the processes of natural selection.

Evolution, as a computational process, uses trial and error methods, creating and testing different variants of genetic combinations. Those that are better adapted to the environmental conditions are more likely to survive and leave offspring. Over time, this dynamic and self-regulating system becomes increasingly complex and sophisticated, creating such wonderful structures as the human brain.

The human brain is the pinnacle of evolutionary development, a tool that has allowed our species to rise to the highest levels in the hierarchy of life. It gives us the opportunity not only to react to the environment but also to actively change it, creating new realities and new opportunities for adaptation. The brain allows us to be aware of ourselves and the world around us, to ask questions and seek answers, to expand the boundaries of our knowledge and influence.

Sensory Sensations as a Result of the Evolutionary Process

Realize this: all the sensory sensations you receive while living your life, the colors and geometry that you see, the sounds that you hear, the hardness or softness of the ground under your feet that you feel, your ability to move in space, your ability to feel that water is wet and mud is viscous, to feel the play of the wind on your skin. Ultimately, you have manipulators in the form of hands. All of this did not come out of thin air, but from the fact that over millions of years, through evolution, genes were derived that carry the most reliable knowledge about the environment.

This knowledge, embodied in your sensory systems, allows you to interact with the environment, receive stimuli, and respond to them. Your eyes perceive light, your ears perceive sound waves, and your skin perceives tactile sensations. All of this is the result of millions of years of adaptive changes that have allowed our ancestors to survive and thrive. Your sensory systems are complex mechanisms designed by evolution to provide you with information about the world around you in a form that can be interpreted and responded to.

And this knowledge is embodied in you precisely because of these genes, which carried reliable information about the environment. You are capable of interactive interaction with the environment, receiving stimuli, and exhibiting an appropriate response. Conversely, you are also able to provide stimuli to the environment, receiving a corresponding response from it.

However, humans do not have genes that would allow them to see the entire spectrum of electromagnetic waves or all the dimensions predicted by string theory. Due to the lack of certain genes, perhaps some things in the universe are fundamentally unknowable to us, and perhaps this is for the better. But we have genes that give us a predisposition to unprecedented abstract thinking, thanks to which we can learn about some phenomena that are inaccessible to us at the level of feelings.

As long as the knowledge contained in these genes is sufficient to implement a survival strategy in their ecological niche, they will continue to exist. And here it becomes obvious that the survival of knowledge itself is fundamental, and not necessarily the gene or any other physical object. It turns out that it is not the physical object that adapts to the niche, but the knowledge that it carries. If the adaptation is successful, this knowledge remains in the niche and affects it. The physical shell of the gene is less important because, during replication, a new copy of the gene is assembled from new components. In addition, knowledge can be successfully transmitted in various physical forms, just as information from a vinyl record can be transferred to magnetic tape and then to a CD.

Perhaps, in the future, people will be able to copy the knowledge contained in their genes onto the most reliable media, replacing their fragile biological essence with something more durable. It would be strange not to consider such people alive. If you do not consider the body a special vessel for the soul, but claim that artificial intelligence is impossible, this is analogous to claiming that the brain is not a physical object. Although all known life is based on replication, in reality, it is built around a single phenomenon—knowledge.

So, our genes are a reflection of the world in which we were able to pass on the genetic code to the next generation. But it is important to understand that we see this world only partially. Our sensory systems are limited, and we are not able to perceive all the richness of physical phenomena occurring in the universe.

This limitation means that our understanding of the world will always be incomplete if we rely only on our senses and genes. Therefore, I propose to move on to the next section, where we will try to understand how physical phenomena and scientific discoveries can help us better understand the world around us. With the help of physics, we can look beyond our biological perception and learn more about what lies beyond the visible and tangible.

CHAPTER 4: THE WORLD BEYOND OUR SENSES

Studying Physical Phenomena

Imagine an infinitely large, completely empty, sterile room, in which there is not a single speck of dust and into which no light penetrates at all. Now imagine a flashlight turned on in this room. It sounds simple, but there is one important detail: if we looked at this room from the side, we would not see either the flashlight itself or the beam of light it emits.

Paradoxically, light is invisible in itself. We only see it when it enters our eyes. We cannot see light that simply passes by without reflecting anything. If there were a second light source in the room, we could see the flashlight, but again not its light. Light rays, even from the most intense source, pass through each other as if they did not exist at all.

Remember this, and now imagine that the flashlight is shining directly into your eyes, and you are gradually flying back. As you move away, the light from the flashlight will appear smaller and smaller, and then turn into a dot altogether. As you move further away, this dot will become dimmer and dimmer. When the distance between your eyes and the flashlight reaches 10 thousand kilometers, the dim dot will disappear altogether. But this is if we talk about the human eye. The eye of a frog is several times more sensitive than the human eye, and this will be enough for the experiment to give a different result.

If the observer is a frog, and it moves away from the flashlight, the moment when it completely loses sight of it will never come. Instead, the frog will begin to observe an amazing phenomenon: it will see that the flashlight has begun to flash at uneven intervals. These intervals will increase as you move away. At a distance of 100 million kilometers, the frog will see an average of only one flash of light per day, but this flash will be as bright as any other.

This flicker indicates that there is a limit to the uniform stretching of light. Each flash that the frog sees is caused by a photon hitting the retina of its eye. Moving away from the flashlight, the photons themselves do not become weaker, but the beam of light weakens due to the fact that the free space between the photons increases. When the frog sees nothing, it is not because the light entering its eyes is too weak, but because no light enters its eyes at all.

This property of appearing only in the form of separate pieces of discrete sizes is called quantumness. Each individual photon is the same discrete piece of light that we can perceive.

Quantum theory got its name precisely from this property, which it ascribes to all measurable physical quantities. There are no measurable continuous quantities in physics. Everything that surrounds us and seems continuous is not really so.

In this section, we will look at the basic principles of quantum theory and try to understand how they help us unlock the secrets of the universe. This will allow us to see the world in a new light, using physics as a key to a deeper understanding of reality.

Next, let's turn to the book by British theoretical physicist David Deutsch, called "The Fabric of Reality." This is his opus magnum (masterpiece), in which he sets out his deepest views on the nature of the Universe. Deutsch, although a scientist and not a writer, expresses his ideas in a rather confusing way, which can make them difficult to understand. However, let's try to understand his thoughts and use them to build a holistic picture of our world.

Many Worlds

Let's be honest, no one takes the idea of the multiverse seriously. Most don't even know what this idea means within physics. In fact, it means a lot and is described differently in different physical concepts. Many of these concepts have a right to exist. We will talk about the many-worlds interpretation of quantum mechanics, which, being a direct and one of the simplest consequences, presents us with a completely new view of the world.

This view is so new that when scientists began to gradually discover the laws of quantum mechanics, they tried to explain what they saw. As a result, they were so disappointed that today among specialists in quantum theory it is considered bad form to talk about anything other than equations. An attempt to explain what these equations say is immediately recorded as empty chatter.

David Deutsch, a British theoretical physicist, in his book "The Fabric of Reality" is contemptuous of such an instrumental approach to physical theory. From his point of view, the desire not only to predict and use but also to understand is absolutely normal.

The Most Tested Theory

Quantum mechanics is a triumph of human intellect, a theory that has passed the most rigorous tests and become the foundation of our technological civilization. If you are reading this book online, this is only possible due to a deep understanding of the quantum laws that govern the behavior of electrons in semiconductors, which are the basis of modern electronics. With all this, I repeat, we do not understand what it is.

However, and this may seem paradoxical, quantum mechanics remains one of the biggest mysteries of science. If Einstein's general theory of relativity gives us a vivid image of curved space-time, then quantum mechanics does not offer such an intuitive understanding. It opens before us a world where particles can be in several places at the same time, where reality depends on the act of observation, where probability replaces accurate predictions.

We have a huge arsenal of mathematical tools that allow us to make incredibly accurate calculations and predictions, but what is behind these formulas? What is quantum reality? There is no single answer to this question. We only have different interpretations, each of which offers its own view of the quantum world, and none of them is universally accepted.

The Main Problem:

Quantum mechanics is an extremely successful theory that allows us to make accurate predictions about the behavior of microscopic particles. However, it does not give us an intuitive understanding of what is really happening. This leads to the fact that physicists are often limited to only mathematical equations, without trying to explain their meaning.

The Double-Slit Experiment: A Window into the Quantum World

One of the most famous and mysterious experiments in quantum mechanics is the double-slit experiment. It clearly demonstrates the strange nature of the quantum world and challenges our usual ideas about reality.

The Essence of the Experiment:

Imagine a thin plate with two narrow slits. A screen is placed behind the plate. If a beam of light (or any other particles, such as electrons) is directed at the slits, then an interference pattern will appear on the screen - an alternation of light and dark stripes.

If light consists of particles, then we would expect to see two separate stripes on the screen, corresponding to each slit. But instead, we see an interference pattern, which is characteristic of waves. It is as if each particle of light passes through both slits simultaneously and interferes with itself.

To put it simply, imagine that we are shining a flashlight through two slits onto a wall. We expect to see two stripes of light, but instead, we see many stripes, as if the light has split into several beams and they interfere with each other.

If we close one slit, then these stripes will disappear, and we will see only one stripe of light. But if we open two more slits, the stripes will reappear, but in a different order. It is as if light passes through all the slits at the same time and creates a complex pattern.

Even if we shine a flashlight very weakly to emit one ray of light at a time, the pattern will still appear. This means that each ray of light seems to know about other rays and interacts with them, even though we do not see it.

Let's approach this issue critically. We found that when one photon passes through this apparatus, it passes through one slit. Then something affects it, causing it to deviate from its trajectory. This influence depends on what other slots are open. That is, objects that affect the photon pass through other slits and behave like photons, but they are invisible.

We can assume that these are some kind of "shadow" photons that can only be detected indirectly, through their effect on real photons.

Each real photon is accompanied by an escort of shadow photons. When a photon passes through one of the slits, some shadow photons pass through the other slit. There are clearly many more shadow photons than real ones. Deutsch and his colleagues concluded that there is no upper limit to the number of shadow photons, but the minimum is a trillion shadow photons per real one.

Follow the thought carefully: what should happen at the microscopic level when shadow photons encounter a light-proof barrier? Of course, they stop. We know this because interference stops when an opaque partition appears in the path of shadow photons.

But why? What stops them? They certainly cannot be absorbed by the real atoms of the partition, because, given the estimated number of shadow photons, the partition would simply evaporate. We would easily fix it in many different ways.

Shadow photons do not interact with real atoms, but the baffle still affects both real and shadow photons. It stops them. But shadow and real photons, in turn, affect the partition in different ways. As far as we know, shadow photons do not affect it at all, but they still stop.

So, along with the existence of a real partition, there is also a shadow one. It's just an inevitable conclusion. Without much effort, we understand that this shadow partition consists of shadow atoms, in which, as we already know, there must be shadow electrons, shadow protons, and shadow neutrons.

To put it simply, imagine that a photon is a ball of light that you throw. When it flies, it seems to bump into other invisible balls from parallel worlds. We do not see these collisions, but they affect the movement of our photon, changing its trajectory.

Could this really be true?

Of course, these are very far-reaching conclusions, and we are dealing with the microcosm. How can we talk about whole parallel universes based only on the interference of photons, quanta of light that do not even have mass?

But we have already obtained interference patterns on fullerene molecules, one of the forms of pure carbon. This is practically a classic object. In other words, particles are grouped into parallel universes.

They are parallel in the sense that within each universe, particles interact with each other in the same way as in our universe. But the influence of each universe on others is very weak and manifests itself through the phenomenon of interference.

Thus, we have derived a chain of inferences that begins with strange shadow patterns and ends with parallel universes. At each stage, we find that the behavior of the objects we observe can only be explained by the presence of invisible objects and their certain properties.

A real photon is tangible, and a shadow photon is only a possible, but not realized, path of a real photon. Quantum theory is about the interaction of the real with the possible.

But why not leave everything as it is? Because, as Deutsch says, the possible cannot interact with the real. Non-existent objects cannot change the trajectory of existing ones. If a photon deviates from its trajectory, something must influence it. It cannot be that a real event (the appearance of a photon) was caused by an imaginary event (what the photon could have done but did not).

As I said, interference is not unique to photons. Quantum theory predicts, and experiments confirm, that interference occurs with any particle. Thus, each real electron must be accompanied by a mass of shadow electrons, each real neutron by a mass of shadow neutrons, and so on.

So, reality is much bigger than it seems, and most of it is invisible. We could call the collection of shadow particles a parallel universe, since shadow particles are affected by real particles only through the phenomenon of interference.

But we can go even further. Shadow particles are separated from each other in the same way as the universe of real particles. In other words, they form not one homogeneous parallel universe, but a huge number of parallel universes, each of which is similar in composition to the real one and obeys the same laws of physics, but differs from others in the arrangement of particles.

The Mystical Disappearance

Imagine an experiment so impressive that it challenges the very concept of reality. We direct a beam of light onto a screen with two narrow slits.

What do we expect to see? Two stripes of light, one opposite each slit. But instead, an intricate pattern of alternating light and dark stripes appears on the screen - an interference pattern (we already know this).

This phenomenon is easy to explain if we imagine light as a wave. Waves, passing through the slits, interfere with each other, creating the observed pattern. But what happens if we try to determine which slit each photon of light passes through?

At this moment, something incredible happens: the interference pattern disappears, and only two stripes remain on the screen. Photons begin to behave like particles, passing strictly through one of the slits, as our brain expected.

The Invisible Observer

It's as if the photons "know" they're being watched and change their behavior. But how is this possible? Maybe the detector somehow affects the photons?

Scientists conducted an experiment by installing a detector near only one of the slits. And even in this case, the interference pattern did not appear when the photons passed through the other slit. The photons behaved as if they "knew" about the possibility of being measured, even if the measurement did not take place.

This phenomenon is called the "collapse of the wave function." A photon, being in a superposition of states (that is, potentially passing through both slits), when measured, "chooses" one of the states and becomes a particle.

The Copenhagen Interpretation

The classical interpretation of quantum mechanics, known as the Copenhagen interpretation, attempts to explain this paradox. According to this interpretation, the world is divided into quantum objects (photons) and classical measuring devices.

A photon, before measurement, exists in the form of a probability wave, being in a superposition of states. But when interacting with a classical device (for example, a detector or a screen), the wave function collapses, and the photon becomes a particle in a certain place.

Questions Without Answers

The Copenhagen interpretation raises many questions and controversies. The division of the world into quantum and classical seems artificial and has no clear justification. In addition, there is a feeling of the mystical influence of the observer on the quantum world.

Some scientists propose alternative interpretations, such as the many-worlds interpretation, in which each measurement leads to the splitting of the universe into many parallel worlds.

The Multiverse and Its Probabilities

If the number of universes is infinite, then there are events so unlikely that they simply do not exist in the multiverse. But if there are an infinite number of universes, then everything that can happen happens, everything that does not contradict the laws of physics. Even the explosion of the Sun due to the random movement of all its atoms to the center. Of course, such an event has a negligible probability, but a branch with such an event exists.

Strictly speaking, in this case, there are an infinite number of universes where the Sun exploded. But this infinity is incomparably smaller than the infinity of universes where the Sun feels fine. The density of these infinities is different, even if their number is infinite.

I don't like using infinity. This concept breaks my brain and it doesn't fit very well with the physical laws where everything is explained by conservation laws, so for example, let's say that each atom has a trillion possible paths. When two atoms interact, the number of combinations of these paths becomes huge, but still finite. This approach avoids the need to operate with infinities while retaining the main idea of the many-worlds interpretation - the splitting of reality into many branches with each interaction.

This approach retains the main idea of the many-worlds interpretation - the splitting of reality into many branches with each interaction - but makes it more understandable and consistent with our intuitive perception of the world.

Many-Worlds Interpretation and Decoherence

The many-worlds interpretation, unlike the Copenhagen one, states that the whole world is quantum. This is more consistent than the Copenhagen quantum-classical world. There is no mystical observer in the many-worlds interpretation that destroys the wave function. Instead, there is a process of constant splitting called decoherence.

In the double-slit experiment, photons fly through the slits in all possible ways at the same time. By measuring a photon, we get entangled with it and split ourselves into all versions of ourselves, having measured the photon in all possible states. Thus, we spread along the branches of probability in the multiverse.

Entropy and Measurement

Measurement is any interaction that is irreversible. The observer does not have to be intelligent for this. The moon really exists, even when no one is looking at it, because the moon is a huge macrosystem in which irreversible interactions are constantly taking place.

By measuring a quantum system that is in superposition, the observer becomes entangled with it and splits into all possible versions of himself, thus finding himself in superposition for another observer. Then the second observer observes the first, gets entangled with him, and also splits into all possible versions of himself, and so on, until the entire universe is involved.

What we observe as the collapse of the wave function is nothing more than our inability to unravel the object and the environment with which it is entangled.

Decoherence and the Collapse of the Wave Function

The many-worlds interpretation offers an alternative view of the collapse of the wave function. Instead of a mystical "observer" that causes the collapse, it introduces the concept of decoherence - the process of losing coherence between different states of a quantum system as a result of interaction with the environment.

In the double-slit experiment, when a photon interacts with a detector, decoherence occurs, and we observe only one of the possible outcomes - a photon that has passed through one of the slits. Other possible outcomes do not disappear, but continue to exist in other branches of reality with which we have lost coherence.

Entropy and Measurement

The process of decoherence is closely related to the concept of entropy - a measure of the disorder of a system. Measurement, as the interaction of a quantum system with the environment, leads to an increase in entropy and, accordingly, to decoherence.

This process is irreversible, which explains why we cannot turn back time and see the interference pattern after the photon has been measured. Information about the initial state of the system is lost due to decoherence.

Quantum Interference and Solids

Quantum interference plays an important role in the formation of solids. If there were no quantum interference, atoms could not form stable structures, and matter would exist only in the form of disparate particles.

Superposition and Isolation

Creating a superposition for macroscopic objects such as the Moon is an extremely difficult task. This requires absolute isolation from any external influences, which is practically impossible to achieve.

Even a single photon reflected from the surface of the Moon can destroy the superposition. Therefore, we cannot observe quantum effects on a macroscopic level in everyday life.

Subjectivity of Time and Space

In our daily lives, we usually assume that there are objective "now" and "here" that are the same for everyone. We live with the feeling that time flows continuously from the past through the present to the future, and that space is a stable three-dimensional stage on which the events of our lives unfold. However, modern scientific theories, in particular quantum mechanics and the theory of relativity, call into question these intuitively understandable ideas.

Quantum mechanics, especially in the many-worlds interpretation, offers a radically different picture of reality. According to this interpretation, each possible outcome of a quantum measurement occurs in a separate universe. Thus, what we perceive as a single stream of events is actually just one of many parallel stories. This means that "now" and "here" can have many different meanings depending on which universe the observer is in.

Each observer has their own "now" and "here," which are determined by their individual experience. This experience can be very different from the experience of other people. For example, during extreme events, such as an accident or intense sports activity, time may seem to slow down or speed up. These subjective changes in the perception of time emphasize that our sense of time is personal and context-dependent.

We perceive time as a flow, but this is just an illusion created by our consciousness. Science suggests that time may be more like a fourth dimension of space than an arrow flying from the past to the future. In space-time, all moments exist simultaneously, like frozen frames of a film. Our consciousness moves from one frame to another, creating the illusion of time movement.

The idea that all moments in time exist simultaneously has profound philosophical implications. This means that our sense of change and the flow of time is subjective, not an objective phenomenon. Perhaps the past, present, and future exist simultaneously, and we simply perceive them in turn due to the peculiarities of our consciousness.

Thus, modern scientific theories undermine our intuitive understanding of time and space. They show that these concepts can be much more complex and subjective than we used to think. This opens up new horizons for understanding the nature of reality and our place in it.

CHAPTER 5: Time as an Illusion

Subjective Perception of Time

In 1972, French geologist Michel Siffre spent six months in a cave in Texas, studying the effects of prolonged isolation on the human body. The conditions were harsh: constant temperature, no sunlight, and no clocks.

Siffre lived in a free rhythm, independently determining his sleep and wake cycles. However, his perception of time began to change. Initially, his subjective days stretched to 25-26 hours, and then reached an astonishing 48 hours, twice the norm.

When the experiment ended, Siffre was surprised to learn that he had spent 179 days in the cave, not 151 as he had believed. His internal sense of time had slowed down so much that he was off by 28 days.

Siffre's case is not unique. Other isolation experiments have also shown that people tend to lose their sense of time.

- In 1988, Veronica Le Guen spent 111 days in a cave in France, but believed only 42 days had passed.
- In 1989, Italian designer Stefania Follini spent 4 months in a deep cave, but believed she had only been there for 2 months.
- In 1993, an Italian sociologist lived in an underground cave for a whole year, but upon emerging, he believed it was summer, when in fact it was winter.

The cases above clearly demonstrate that our subjective perception of time is not objective. You might argue that, like any subjective experience, subjective time is merely a product of the brain's activity and does not exist outside of it.

Of course, like most subjective feelings, the sense of time is also subject to illusions and distortions. But wait! When we look around, we see a whole picture of the world created by the brain. However, when it comes to a subjective phenomenon, such as color, we know (as much as possible) that color has a physical equivalent – the wavelength of electromagnetic radiation. That is, the brain, albeit subjectively, interprets an objective phenomenon that really exists in nature.

But what does the brain interpret when we talk about the feeling of the passage of time? What is the real equivalent of this feeling? If you think this is a silly question with an obvious answer, then you don't even know the little that science knows about time today. Because what you are feeling right now, the feeling of the present moment, is one of the main unsolved scientific mysteries.

The Objectivity of Time

What is time objectively? There are also big problems with this. Even St. Augustine, a philosopher and theologian, wrote more than a thousand and a half years ago: "If no one asks me about it, I know what time is. But if I wanted to explain it to the questioner, I don't know." Physicist Richard Feynman said: "It would be nice if we came to terms with the fact that time is one of those things that we probably won't be able to define."

Indeed, it is not how we define time that matters, but how we measure it. Therefore, the working definition literally sounds like this: "Time is what clocks show."

But if you still try to answer this question in detail, most people will intuitively come to the conclusion that time is a sequence of events that flows from the past to the future through the present. In this case, only the present moment is real, because every moment that goes into the past immediately disappears, and the future has yet to happen and does not yet exist. In general, the past and the future are just our abstractions.

Einstein's Train and the Relativity of Simultaneity

Imagine a train moving at great speed. In the center of the train is a person with two pistols, ready to fire at the windows at the beginning and end of the car. These pistols are so powerful that the bullets they fire travel at a speed approaching the speed of light.

When the person fires, something amazing actually happens. For an observer on the train, both windows will break simultaneously. However, for an observer on the platform, things will be a little different. The bullet fired back will reach the rear window and break it first, as both are moving in the same direction. But the bullet moving forward has to overcome the movement of the train, so it will reach the front window a little later. This is because for the observer on the platform, the bullet moving forward has a higher speed than the bullet moving backward, and time will "slow down" for it so as not to violate the speed limit of light.

Thus, from the point of view of the observer on the platform, the rear window will break first, and then the front window. This phenomenon, which contradicts our intuitions, is the result of special effects known as the effects of special relativity.

This phenomenon is called the relativity of simultaneity. It suggests that there are two different realities in our world. In one reality, both windows are either intact or broken simultaneously, while in the other there is a moment when the rear window is already broken and the front window is not yet.

These are not two different universes, both people are in the same universe. But they won't agree on when the windows broke.

No matter how strange this situation may seem, special relativity states that both realities are equally real. There is no absolute "now", the simultaneity of events depends on the observer's frame of reference.

This paradox challenges our intuitive understanding of time and space. It shows that reality is much more complex and amazing than we used to think.

Two Equal Realities

Special relativity claims that both realities described in the thought experiment with Einstein's train are absolutely equal. Although this may seem strange and illogical, this is the picture of the world that the laws of physics paint for us.

Often at this point the discussion stops, and we simply accept this fact as a given. But what should the universe look like in which this is possible? How can contradictory statements be true at the same time?

Understanding that both realities in a thought experiment are equal requires a deep refinement of our understanding of the nature of time, space, and motion. This task before us requires not only the ability to accept intuitively non-obvious physical phenomena, but also the ability to understand how these phenomena interact on the vast scales of the cosmos.

In this context, we can imagine the universe as a complex system in which every object and every event affects others. According to the theory of relativity, each point of space-time has its own independent history, and observers moving relative to each other may have different ideas about what is happening simultaneously.

Thus, the equality of both realities lies in the fact that neither of them is more objective or true than the other. Both realities exist and have their own laws that reflect the physical principles we observe in our universe.

A Model of the Universe That Defies Intuition

For over a century, scientists have been trying to create a model of the universe that would be consistent with the laws of physics and explain the paradox of simultaneity. And such a model exists, although it is far from our everyday intuition.

This model is accepted by many physicists and philosophers because it is supported by the laws of physics. However, it is not very comforting, because the picture it paints is not very pleasant and encouraging.

The most complex mechanism in the universe - our brain - did not evolve to understand the nature of time. However, the contradictory picture of the universe that physics reveals to us has probably influenced the very construction of the human brain.

As we will see later, the structure of our brain, along with the laws of physics, in a certain way hints at what reality and time are.

The book by American neurobiologist Dean Buonomano "Your Brain is a Time Machine" explores how the human brain encodes time. Buonomano, one of the first neurobiologists to devote a significant portion of his career to this issue, studied the work of various scientists to understand how our brain creates a sense of the flow of time.

Science always strives to separate the experiment from the experimenter in order to achieve maximum objectivity. However, in the case of the study of time, Dean Buonomano tries to do the opposite. He seeks to combine the subjective experience of the perception of time with objective scientific data.

The Illusion of Time

For more than a century, scientists have been trying to find at least something in the physical world that could be called the flow of time. So far unsuccessfully. Therefore, Buonomano, like many other scientists before him, comes to the conclusion that our perception of the flow of time can only be an illusion.

Einstein, explaining his theory of relativity, said that time is relative and depends on the movement of the observer. For example, for a person moving at high speed, time will go slower than for a person at rest.

Buonomano believes that our sense of the flow of time is the result of the brain's work, and not a reflection of some objective reality. He compares this feeling with other subjective feelings, such as color or taste, which are also the result of the brain's interpretation of signals from the senses.

Internal Clock and Predicting the Future

Despite the distortion of the perception of time in conditions of isolation, our body has an internal clock that helps us navigate in time. We intuitively sense when the green traffic light will turn on or when the TV commercial will end.

According to Dean Buonomano, the mechanisms for determining time are built into the operating systems of the brain at the most basic level - at the level of neurons, synapses and their networks. Therefore, it makes no sense to look for a separate part of the brain responsible for the perception of time, since most neural networks are involved in this process in one way or another.

In the broadest sense, the brain can be called a time machine. Of course, not in the sense of time travel, but in the sense of working with time. For hundreds of millions of years, animals have developed the ability to predict the future. Predators learned to predict the behavior of their prey, and prey - the behavior of predators. They all tried to predict the behavior of potential partners.

Some animals prepare for the future by storing food, building nests, and so on. Life on Earth anticipates the change of seasons, day and night. Those who did not cope with this task did not survive and did not leave offspring.

Automatic Prediction of the Future

Whether we realize it or not, our brain is constantly trying to predict what is about to happen. These short-term predictions, approximately a few seconds in advance, are made automatically and unconsciously. For example, if you drop a ball from a table, we automatically make a movement to catch it when it bounces off the floor. But we react quite differently if a piece of cake falls from the table.

People and other animals are constantly trying to make predictions for a variety of time periods. A cat, having got into a new house, tensely builds a map of the area in its head, sniffs everything around, preparing for what might happen not only in a few seconds, but also in a few minutes or even hours.

A wolf, stopping to pick up some signs, sounds and smells, looks for clues that will help him identify potential enemies, prey or a mate.

Even pollinating birds are able to measure the time elapsed since their last visit to a particular flower so that the nectar has time to accumulate by the next time.

Internal Clock and Predicting the Future

Almost all manifestations of life, from the ability to hit a moving target with a spear, to understand when to laugh at the end of a joke or play Beethoven's "Moonlight Sonata", to the ability to regulate the daily cycle of sleep and wakefulness or the monthly menstrual cycle - all this requires the ability to determine time.

The brain not only counts the seconds, hours and days of our lives, but also recognizes and creates temporal images, such as musical rhythms and precise sequences of movements that allow gymnasts to perform acrobatic stunts. Our natural desire to clap our hands, snap our fingers or nod our heads to the beat of the music.

Your brain looks ahead a few hundred milliseconds, predicts the next beat, and syncs your actions to it. If you want to understand how deeply embedded this is in us, try to break the rhythm of the music and, for example, snap your fingers out of time. To do this, you will have to focus all your attention, while maintaining the rhythm of the music requires almost no concentration.

The Brain Creates the Flow of Time

Not only prediction, but also the very feeling of the flow of time, the continuity of the present moment, is the creation of our brain. This is easy to verify with a simple experiment.

Ask someone to stand in front of you and start looking at your left and right eyes alternately. You will notice that the person's eyes are moving and this movement takes some time.

Now go to the mirror and try to do the same, looking at the left or right eye of your reflection. You will find that your reflection does not blink at all, does not even try. This is because the brain simply cuts out this moment and all the moments that occur when you move your eyes from one object to another. We don't even notice it, the picture seems continuous to us.

The same thing happens when blinking. The brain stitches together frames before and after closing the eyelids. You can say that these are trifles, but, for example, at a speed of 100 km / h, a car travels about 5 meters during a blink. 5 meters that simply do not exist for you.

It is clear why this is happening. We did not evolve to make decisions in such conditions, so it is so dangerous to drive in the city. Formula 1 racers, where speeds exceed 350 km/h, generally learn to blink only in certain sections of the track.

In general, during the day we run about an hour of cut material. So, the illusion of the continuity of reality is the merit of the brain. The only thing we live directly is the eternal "now".

The Eternal "Now" of Clive Wearing

In reality, we cannot be at any other moment than in the "now". And at the same time, when trying to catch this very "now", it immediately slips away.

But not for everyone. There is one man with a rare specific brain injury, due to which his present moment about 40 years ago in some sense still stopped. This is a British music critic Clive Wearing, who, after a serious infectious disease and damage to the hippocampus, the brain completely lost the ability to create strong new long-term memories.

This is one of the most severe cases of amnesia in the world. His memory of events lasts from 7 to 30 seconds. His whole life consists in the fact that on average every 20 seconds he seems to "wake up", restarting his consciousness after the expiration of his short-term memory.

Every few tens of seconds, for almost 40 years now, it seems to him that he has just come out of a coma. If he is involved in a conversation longer than a few sentences, he is advised to keep a personal diary, which he does.

But if you look inside, we will see a picture that causes horror. Page after page, the entries look like this:

"8:30 am. Now I'm really fully awake."

Then Waring crosses out this line and writes:

"9:06 am. Now I'm definitely awake."

Crossed out again:

"9:34 am. Now I'm most definitely truly awake."

Examining his diaries, we see that at some point he begins to write the time in large numbers, with such pressure, as if trying to mark himself in the continuum, to get on the time train.

It's like a small death every few tens of seconds, which he is so desperately trying to fight. The inscription in huge letters: "I'M ALIVE!" And each time he did not know how and by whom the previous entries were made, although he recognized his handwriting.

Since Waring cannot understand where he is or how he got here, the only possible explanation for his brain is that he just woke up. An endless loop of one single moment.

In a 2005 documentary, Waring answers similar questions:

"You are the first people I have seen. You three: two men and one lady. The first people I saw since I got sick. There is no difference between day and night. No thoughts at all, no dreams..."

The story of Clive Wearing is tragic, although it does not prove, but nevertheless suggests that, perhaps, for people the moment that we call "now" is associated with short-term memory and does not occur in one moment, but as if in jerks, having a certain duration in time. That is, the conscious feeling of the present can be compared to a note rather than a freeze frame of a film.

Matter and Consciousness

Neuropsychologist, linguist, and Harvard University Professor Emeritus of Psychology Steven Pinker noted: "Matter is distributed in space, but consciousness exists in time." (Remember this quote, it will be useful in the following sections). This statement is as obvious as "I think, therefore I am."

However, the question arises: how densely is consciousness distributed in time? How short a moment can we capture?

The Influence of Psychoactive Substances on the Perception of Time

Our sense of time changes dramatically under the influence of psychoactive drugs. One of the founders of modern psychology, William James, wrote: "When intoxicated with hashish, there is an interesting feeling of stretching time. We begin to speak a phrase, but when we reach the end, it seems that we started talking an eternity ago."

The active component of hashish or marijuana, tetrahydrocannabinol (THC), indeed, according to experimental data, causes a feeling of slowing down or stopping external time. After using it, people estimated the time interval of one minute as 42 seconds.

But the change in the perception of time occurs not only under the influence of substances. We have often heard, and some have even experienced, the effect of slowing down time during severe emotional distress or life-threatening situations.

The reasons for such a distortion of time are not fully understood. There are several hypotheses:

- Overclocking the brain. By analogy with overclocking a processor, Buonomano suggests that the brain can briefly increase its efficiency by 10-20%.
- Hyper memory. People perceive events in slow motion not at the time of the event, but later, remembering it. During the "fight or flight" response, the brain can increase the temporal and spatial separation of memory. Thus, in retrospect, it seems that everything happened more slowly.
- Subjective distortion of time. The author of the book, having got into a car accident, felt that time had slowed down. However, the video of the accident showed that everything was happening at normal speed. This confirms that our perception of time can be distorted at such moments.

Meta Illusion: The Illusion of Time

Buonomano proposes a third, most interesting hypothesis called "meta-illusion". To understand its essence, try to touch an object with your hand, such as a wall, table, or phone, and observe your feelings. Doesn't it seem strange to you that, although the formation of the sensation of an object occurs in the brain, we do not feel it in our head, but literally transfer it to a specific point in space?

Buonomano writes that one of the deepest subjective feelings of a person is that our fingers, hands, feet, our whole body belongs to us. And all this is one big illusion.

Phantom Limbs and the Illusion of Body Ownership

You've probably heard of phantom limb syndrome. Some people, after amputation of an arm or leg, continue to feel it as clearly as most of us feel real limbs. This phenomenon suggests that the brain is working so hard to create in us a sense of ownership of the bones, muscles, and nerves that make up our limbs that it continues to maintain this illusion despite the disappearance of the limbs themselves.

If you hit your finger with a hammer, your brain will project the sensation of pain into a specific area of space - into your finger. But if you put an artificial hand next to your hand, the brain can change the perception in such a way that you will feel your hand where the artificial hand is, as if the brain agrees to consider the artificial hand yours. This is the so-called rubber hand illusion.

Based on this example, Buonomano suggests that if our brain builds such persistent spatial mirages, then why shouldn't it build temporal ones?

What we call the flow of time may turn out to be an illusion, and so the name of the hypothesis "meta-illusion" means that the slowing down of time is an illusion of illusion.

On YouTube, you can choose the playback speed of the video. You can speed up or slow down the video twice and still perceive the information well. Buonomano writes that our normal sense of time is a mental construct that can have different speed settings.

You can verify this by watching the video at double speed for 5 minutes, and then turning on the normal speed. You will be surprised how slow the usual passage of time will seem.

Buonomano argues that the speed of our perception of time is not a static illusion. In fact, we are constantly using our ability to compress and stretch time.

For example, you can say any phrase in your mind much faster than with your lips and tongue. The same applies to tying shoelaces, getting up from the couch, and any other actions.

His book gives several examples of time distortion in life-threatening situations:

- A 20-year-old race car driver who crashed at 250 km/h says it all happened very slowly and he felt like he was playing on stage, watching himself from the side.
- A 21-year-old boy who fell from a height of 10 meters also felt that time slowed down and he could watch his fall as if from the side.
- A World War II soldier whose car was blown up by a mine says that time seemed to stop and he existed only in thoughts.

As you can see, in critical situations, not only the perception of time changes, but also the perception of space. Many people at such moments observe what is happening, as if from the outside.

Buonomano writes that in any other context, the above statements would seem like hallucinations or impaired consciousness. Perhaps the sudden release of endogenous opioids that occurs in such situations is the cause of such a distortion of perception.

Fundamental Unit of Time

Is there a fundamental unit of time that cannot be divided into an even smaller one? Clocks are the most accurate instrument we have ever created, but even the most modern atomic clocks have discrepancies in readings if they are placed at different heights.

Richard Feynman once said that due to the effects of the theory of relativity, the Earth's core should be noticeably younger than its crust. Recent calculations have shown that over the entire existence of the Earth, the difference between the core and the crust has accumulated by about 2.5 years.

Science does not yet have an answer to the question of whether time is discrete or continuous. Many experts believe that the existence of individual moments would lead to paradoxes, such as Zeno's paradox of dichotomy.

This paradox goes like this: To travel a path, you must first travel half the path, and to travel half the path, you must first travel half the half, and so on ad infinitum.

Andromeda Paradox

The laws of physics are symmetrical with respect to time, that is, they do not attach special importance to its direction. The past, present and future are equivalent to each other. This means that "now" on the time scale is the same as "here" in space.

But Einstein's theory of relativity complicates this picture. The example with the train shows that each observer, depending on the speed and direction of movement, has his own independent concept of the present moment.

Roger Penrose, in his book The Emperor's New Mind, gives a thought experiment that forces us to reconsider our ideas about reality. He shows that even at very small relative speeds, changes in chronology become colossal if two points are at great distances from each other.

For example, two pedestrians slowly passing each other on the street will not see the difference between the events taking place around them. But if at the moment of their meeting we move to the Andromeda galaxy, then the events simultaneous for them will actually be several days apart from each other.

This means that there are an infinite number of planes of simultaneity passing through any point in space-time. For each point in space, there are different sets of simultaneous events.

Even the slightest movement of your head changes the perceived present moment for you. In the universe, everything becomes even more absurd when you realize that the spaces of present moments are different for your head, arms, legs, and body.

What Does the Universe Look Like?

If we could fly beyond the cosmos and look back at it, from the perspective of special relativity, we would see an unchanging four-dimensional block where time exists as another spatial coordinate.

Within this model, the block universe model, talking about "now" is the same as talking about "here," because any present moment is real and exists on one of the cross-sections of this block.

The idea of a block universe is not just an appealing metaphysical theory, but a well-established scientific fact. Interestingly, when Einstein first published his paper on special relativity, he did not claim that time should be considered as the fourth dimension of a block universe. It was his teacher in Zurich, Hermann Minkowski, who first drew these amazing conclusions about the relationship between space and time.

Minkowski presented Einstein's theory in geometric form, combining space and time into a single four-dimensional continuum – spacetime. In this spacetime, each event has its own coordinates: three spatial and one temporal.

For illustration, one can imagine a simplified two-dimensional spacetime, where one axis corresponds to time and the other to space. In this representation, the plane of simultaneity is a line that passes through a certain moment in time and connects all events that occur simultaneously from the point of view of a given observer.

Fatalism

If we imagine that time is not a flowing river, but a frozen block where all events of the past, present, and future are already determined, then the question of free will arises. Are our choices truly free, or are they just an illusion caused by our subjective perception of time?

The philosophical concept that asserts that all events in the world are predetermined and inevitable is called fatalism. Within the framework of a block universe, where the future already exists, fatalism may seem like a logical conclusion.

Free Will in a Block Universe

However, even in a block universe, there is a possibility for different versions of the future. This is due to the fact that we cannot know all the details of the initial state of the Universe, and therefore cannot accurately predict all future events.

In addition, quantum mechanics introduces an element of randomness into physical processes. This means that even if we know the initial state of the system, we cannot predict its future state with absolute accuracy.

Therefore, even in a block universe, there is a possibility for different versions of the future, and our choices can influence which of these options is realized. However, our sense of free will may be an illusion caused by the fact that we do not know all the details of the initial state of the Universe and cannot accurately predict the future.

Physical Determinism and Free Will

Physical determinism, the idea that all events in the world are determined by previous events and the laws of physics, does not necessarily contradict free will. We can consider free will as the ability to act according to our desires and beliefs, even if these desires and beliefs themselves are determined by physical processes.

The question of free will is closely related to the question of the nature of time. If time is just an illusion, then can we talk about freedom of choice? And if time is real and has direction, then can we change our future?

Spatialization of Time by the Brain

Against this backdrop, it becomes interesting that humans likely developed the ability to understand the concept of time using the same mechanisms designed for comprehending space. In other words, at a basic level, the brain may not distinguish between space and time.

The renowned Swiss psychologist Jean Piaget sought parallels between psychology and physics. He revolutionized the field of developmental psychology by explaining the mechanisms of children's cognition of abstract concepts such as quantity, space, and time.

Piaget likely believed in the existence of a deep connection between children's innate understanding of the relativity of time and the relativity of time in Einstein's theory. To understand how time is reflected in children's minds, he asked them to perform various simple tasks.

In one such task, Piaget used two snakes crawling along parallel paths for several seconds. For example, a blue and a yellow snake would start moving from the same initial position at the same moment in time and stop simultaneously. But the blue snake would move further because it crawled faster.

Children aged 5-6 mistakenly reported that the snake that crawled a greater distance stopped later. That is, the parallel with the theory of relativity is as follows: children intuitively understand that for an object moving at a higher speed, time stretches.

Mental Timeline

How do adults represent chronology? Imagine the years 2021, 2022, and 2023 in chronological order. Most likely, you arranged them from left to right. This seems natural, but why is it so? After all, the time scale can be imagined in any way.

If we use space to denote time, why not from right to left or from bottom to top? Wouldn't that be more like moving forward in time? But no, people most often imagine the time scale from left to right.

There are experiments that confirm the existence of a mental timeline directed from left to right. For example, in a study where participants had to compare the duration of notes with a certain standard, people coped with the task faster and better if they could use the index finger of their left hand to indicate a short interval and the index finger of their right hand to indicate a long interval.

We constantly use spatial metaphors to describe time: "running ahead," "looking back," "short time," "long time," and so on. Metaphors from the field of space are often used to describe time, and very rarely vice versa.

Intertwining of Space and Time in the Brain

Although we do not yet fully understand how neurons in the hippocampus or other brain regions measure, reproduce, and store information about the magnitude of spatial and temporal parameters, based on philological, psychophysical, and neurophysiological data, we can conclude that space and time are intertwined in our neural circuits.

Movement in Time and the Geometry of Space

Does the geometry of space change when moving in time? Although time is very different from spatial dimensions, when we move, we see how the geometry of space changes: objects appear larger when we approach them and smaller when we move away.

Changes also occur when moving in time, although they are not so obvious. Objects are compressed along the direction of movement. For example, at a speed of 60 km/h, a car 5 meters long appears 8 micrometers shorter.

At speeds close to the speed of light, this effect becomes more significant. If the Saturn V rocket could reach a speed of 299,992,457 m/s, the diameter of the Moon in the direction of the rocket's movement would shrink from 3474 km to 284 m.

Subjective and Objective Perception of Time

We have discussed the concept of time from the point of view of our personal perception and from the point of view of physics. Now let's try to combine these two views and get a holistic picture of nature. But this is precisely what is impossible, and this is one of the main mysteries of the universe.

Of all the obstacles to a deep understanding of life, no problem is as terrible as the problem of time. How to explain time? No way, if you don't explain life. How to explain life? No way, if you don't explain time. Revealing the deep and hidden connection between time and life is a matter for the future.

People and all living beings can move along spatial axes in both directions, but movement along the time axis always occurs only in one direction. At least, people know this from their own conscious experience. We can regulate the speed of movement in time, but not the direction. For us, time always moves only forward and never backward.

At the same time, the fundamental laws of physics say nothing about why time seems to move forward for us. The equations of Newton, Einstein, Maxwell, and Schrödinger do not depend on whether events develop in forward or reverse order. They do not have any specific present moment in time.

Despite all these compelling arguments in favor of the fact that we live in a block universe, we have to admit that the laws of physics cannot explain the most important human observation, which is that the present moment is different from all other moments and that time passes.

The Problem of the Flow of Time

Einstein, despite adhering to the concept of a block universe, was also concerned about the discrepancy between our feelings and the modern understanding of the laws of physics. He recognized that the experience of the present means something special for a person, fundamentally different from the past and the future.

This experience cannot be explained by science, and for Einstein, it was a reason for a painful but inevitable retreat.

Roger Penrose, after describing a thought experiment with Andromeda, notes that according to the special theory of relativity, such a concept as "now" does not really exist. The best approximation to it would be the space of simultaneous events of the observer in spacetime. Penrose compares the Universe to a vinyl record, and our consciousness to the needle of a record player.

The Illusion of the Flow of Time

The discrepancy between the idea of a block universe and the feeling of the flow of time is such a deep problem that many physicists and philosophers consider the only way to solve it is to recognize the feeling of the flow of time as an illusion.

Theoretical physicist Paul Davies writes: "The apparent feeling of movement or flow of time, perhaps acquired through the back door of thinking, is the deepest mystery. Is it related to quantum processes in the brain, reflects the objective property of time in our real world of material objects that we simply cannot detect, or will the flow of time ultimately turn out to be exclusively a mental construct, an illusion or a mistake of consciousness?"

The feeling of the flow of time is indeed a mental construct, at least because we perceive the world around us from inside our heads. Vision, as well as sounds and smells, are the same mental construct. These are illusions in the sense that they do not exist in the outside world, but they have adaptive meaning because they correlate with real physical phenomena: the length of an electromagnetic wave, a certain set of sound waves, or the chemical structure of molecules.

In the objective world, there is no blue color, blue is an illusion caused by electromagnetic radiation with a wavelength of 470 nm. In the objective world, there are no unpleasant smells, but there are, for example, sulfur molecules that the brain interprets as the smell of rotten products.

Any such illusion has adaptive meaning because it strictly correlates with real physical phenomena.

A More Fundamental Level of Reality

Theoretical physicist Brian Greene also tries to explain the feeling of the flow of time within the framework of the block universe, comparing each moment in spacetime to a frame of a film. However, many believe that this is not an explanation, but rather an attempt to avoid answering.

Perhaps the struggle of different ideas will help us better understand the nature of time. Or maybe it is even more unexplored and strange than it seems to us at the moment.

A book titled "Nonlocality" by science journalist George Musser was recently published. It combines evidence that there is a more fundamental level of reality than we think, and spacetime is just a derivative of it.

Quoting Professor of New York University and one of the world's leading philosophers of physics Tim Maudlin: "The world is not just a collection of separately existing localized objects connected externally only by space and time. Something deeper, more mysterious holds together the fabric of the universe. We have just reached the point in the development of physics where we can begin to speculate about what it could be."

CHAPTER 6: THE NATURE OF SPACE

Elusive Space

Space is something we take for granted. We live in it, move through it, but can we see or touch it? In fact, space as a physical phenomenon is not an observable object. We can point to objects located in space, to their interactions, but not to space itself.

Waving our hand in the air, we might say that this emptiness is space. But this is just an illusion. Space is not emptiness; it has its own properties and affects matter.

Space is a fundamental concept in physics. All physics studies how objects move in space, and space determines almost all quantities that physics deals with: distance, size, shape, position, speed, direction.

However, some scientific works in the field of cutting-edge physics suggest that what we call space is actually a very suspicious thing. The space between your eyes and the book hides a big mystery.

Theoretical physicists like Max Tegmark, David Gross, and Nathan Seiberg express doubts about the fundamentality of spacetime. They believe that these are just approximate concepts that will soon be replaced by something more elegant.

Nathan Seiberg even argues that space and time are illusions, primitive concepts that will soon be replaced by something more complex. He compares space to the canvas of a painting, which can be removed, but the objects painted on the canvas will remain.

But if spacetime is not fundamental, then what is physics about? After all, all physics studies what happens in space and time. If there is no spacetime, then what is physics about?

How Our Senses Deceive Us

After reading Donald Hoffman's book "The Case Against Reality: Why Evolution Hid the Truth from Our Eyes," I discovered that it contains incredible and implausible things for traditional perception. The author is a serious cognitive scientist who uses mathematical models and puts forward testable hypotheses. For example, on Lex Fridman's channel, the podcast with Hoffman is the most popular in the entire history of the channel. Hoffman's ideas are bold for traditional understanding, but I like that because they make us reconsider our established ideas about reality. They open up new horizons for research and allow us to understand more deeply how our brain and senses interact with the outside world. Hoffman proposes to see the world not just as an objective reality, but as a complex system where our perception is just a tool created for our survival. This makes us think about the fundamental aspects of existence and how we can use this knowledge for the development of science and technology.

This becomes even more exciting after reading George Musser's book "Spooky Action at a Distance," where similar topics are explored from a different perspective. Recommendations from such prominent scientists as Frank Wilczek and Mario Livio add weight to these ideas and confirm their importance in the current scientific discourse.

Computer Modeling of Evolution

Donald Hoffman relies heavily on computer modeling methods, such as simulating the process of evolution. The results of these calculations speak of such counterintuitive things that it is hard to believe them.

For example, Hoffman argues that our consciousness is not a product of evolution, but on the contrary, consciousness is a fundamental property of reality, and it is it that creates the illusion of space and time.

All this leads us to an interesting conclusion: our perception of reality, including time and space, does not necessarily have to reflect the objective truth. Instead, it is shaped by evolution, which strives for the maximum adaptability of the organism to the environment.

This idea can be expressed by the theorem "Fitness beats truth." Our brain does not strive for an absolutely accurate reflection of reality, but rather creates a simplified model that allows us to effectively interact with the world and survive. (Which I mentioned in previous sections)

For example, when we open our eyes, billions of neurons and trillions of synapses are activated. About a third of the cerebral cortex, our most developed computing power, is involved in the process of vision.

This is not exactly what you would expect if vision is just something like shooting a video. After all, cameras appeared long before the era of computers. So what does the brain compute when we look?

Let's start with a creature that, in a sense, understands visible space much better than we do. For it, people are just dots moving on a plane. It's a herring gull.

How do you think seagulls perceive the world around them? It can be assumed that since gulls fly, vision is the most important tool of perception for them. And a person receives almost 90% of information about the world around him through sight. So, we and the gulls perceive reality plus or minus the same, right?

It sounds logical, but the correct answer is: we have no idea what the world of this bird looks like.

Nicholas Tinbergen's Research

Imagine an object that can be described as a long red rod with three white rings. But if you were a newly hatched herring gull chick, you would see your mother instead.

In the 1950s, biologist and Nobel laureate Nicholas Tinbergen conducted research that is fully described in his book "The World of the Herring Gull." Tinbergen tried to understand how newly hatched chicks always unmistakably recognize their mother and do not confuse her with other objects. It is important for the chick to recognize its mother, because in order to get food, it needs to peck at her beak, after which she will pass partially digested food to it through her open mouth.

Tinbergen, conducting experiments with mock-ups of gulls, found that the chicks do not distinguish a real mother gull from a mock-up of a head on a stick. They did not even notice the difference if the mock-up was flat or consisted only of a beak.

In the world of a hungry chick, there is no volume or any details, only a conditional shape and color. The color is most likely because the mother gull has a red spot on her beak. So, only very conditionally shape and color.

One might assume that the chicks are simply still almost blind, after all, they have just hatched. Tinbergen also thought so at first, but tests showed that the chicks' eyesight is in perfect order.

In the end, Tinbergen, guided by accumulated experience and understanding, made a long red rod with three white rings and found that the chicks beg for food from this very far from the original model even more persistently than from their real mother.

Different Objects, Same Experience

So, we have a number of completely different physical objects that, nevertheless, cause absolutely the same inner experience in a living being. What is this anyway?

Donald Hoffman argues that there is actually nothing strange about such things, because evolution, no matter what you think, does not promote a true perception of the world.

Hoffman and his colleagues have conducted hundreds of thousands of simulated evolutionary games. In these mathematical simulations, different environments were generated, and three types of organisms fought for resources in each environment:

1. Organisms that saw reality as it is.
2. Organisms that saw only part of reality.

3. Organisms that did not see any reality and had only a basic adaptation mechanism.

The computer calculated the evolution and interaction of these three types of organisms in each environment. And who do you think eventually won the competition for resources?

According to Hoffman, evolution by natural selection methodically eradicates any reliable perception of reality because reliable perception is inefficient.

Imagine that among the gull chicks, one suddenly appears that sees objective reality. One might think that this would significantly increase its chances of survival. But in fact, while it will figure out whether it is its mother, all the food will be eaten by other chicks who react instantly as soon as they see an oblong shape with a red element.

An organism that sees objective reality is always less adapted than an organism of the same complexity that sees only what it needs to survive. Seeing objective reality leads to extinction.

Simplifying Reality for Survival

Evolution hides the unnecessary complexity of the world around us, directing actions in a purely applied direction. Seeing a mother gull in a red rod with three white stripes is beneficial from the point of view of adaptability.

It is clear that the world of a gull, especially an adult one, is not limited to its mother. But, according to Hoffman, any interaction of a gull with the outside world is built through similar simplifying mechanisms.

Now look at any object in your environment. The perception mechanism formed in the course of evolution tells us that the ball is a cube. But we can come up and touch it to make sure.

People, like newly hatched chicks, cannot understand that the white color of the screen is not actually white. The screen only has blue, red, and green LEDs, and when they are mixed, light appears that we perceive as white, but in fact, it is not. In nature, real white light is sunlight, which as a physical entity is very different from what you see now.

But for our perception, the difference is zero, because this defect did not prevent our ancestors from reproducing in any way.

The "Fitness Beats Truth" Theorem

It is important to understand that any feeling of any living being evolved not to reflect objective reality, but only to respond as quickly and efficiently as possible to stimuli necessary for survival, while spending a minimum amount of energy. This applies not only to vision but also to any sense organ.

Hoffman calls this the "Fitness beats truth" theorem because he uses mathematical proof. Of course, it is very difficult to study the limitations of human perception while being human. But in any case, why should we, with our complex senses, believe that we perceive reality close to what it really is?

The more complex the senses become, the less chance they have of revealing any truth about objective reality. Consider, for example, an eye with ten photoreceptors, each of which has two states. Fitness theory states that the probability that such an eye sees reality is at most two in a thousand. For twenty photoreceptors, the probability is two in a million. For forty photoreceptors, it's one in ten billion. The human eye has 130 million photoreceptors, and the probability that it sees objective reality is practically zero.

Immanuel Kant's Critique of Pure Reason

Donald Hoffman's ideas, despite their mathematical justification, may seem dubious. This is not a new concept. More than 200 years ago, in one of the most fundamental works in the history of philosophy called "Critique of Pure Reason," the German philosopher Immanuel Kant expressed similar thoughts.

Kant argued that the objects and phenomena we observe are not at all what exists in reality. To illustrate, imagine a photograph: our perception is a fabric that envelops something. This "something" exists in reality, and Kant calls it the "thing-in-itself." We do not have direct access to this "thing-in-itself," we cannot tear off this fabric because we ourselves are this fabric.

Following this analogy, we do not just passively observe objects, we "feel" them in our minds. However, this experience cannot tell us anything about the real properties of these objects because there can be anything under the fabric: a cube, a box, or even a computer system unit.

According to Kant, when you look at an object and see an apple, this does not mean that an apple exists in the real world. There is "something" that causes you to experience an apple, but you cannot know what this "something" is. This "something" that causes us to experience an apple is generally outside of space and time, because, from Kant's point of view, space and time are not characteristics of the external world, but ways of organizing our experience.

Simply put, space and time for us are not something that we first experienced and then abstracted as an idea. No, this is what we have before any experience, like, for example, the fear of the dark. We experience and abstract the fear of sockets, but we are instinctively afraid of the dark.

In Kant's analogy, space and time are properties of our fabric of perception. You may ask, why do we need the ideas of this ancient philosopher?

Nobel Prize for a Blow to Realism

Who would have thought that one day a significant scientific base could be fitted under these ideas? More recently, in 2022, the Nobel Prize in Physics was awarded to three scientists, in particular for experiments that seem to refute the foundations of realism.

Realism in physics is the assumption that nature is as we know it, exists independently of the measurement process. You may have heard of experiments with entangled particles that look completely independent, but when measuring the state of one particle, the state of the other always becomes opposite, and this happens at infinite speed. It is important that you can choose at what angle to carry out the measurement, and thus you directly influence the result. That is, you set the particle a framework in which it can act, and it adapts.

The trick is that the second particle, even if it is on the other side of the Universe, instantly learns at what angle its companion was measured and instantly takes on the opposite value, as if there is no distance between them.

Many scientists believed that there was no instant connection, they say, the particles are simplified options from the set: if you took one and saw that it is right, then the second will definitely be left. But the same Nobel Prize was awarded, in particular, for experimental confirmation of the violation of Bell's inequalities. Translating into understandable language, this means that if the particles were gloves, then none of them would be right or left until they were measured.

In summary, firstly, particles do not have properties until they are measured, and secondly, when one particle is measured, the other learns about it instantly. And physicists note that this idea is closer to magic than anything they have seen before.

George Musser, in his book "Spooky Action at a Distance," explains that quantum entanglement, which means the nonlocality of the world, worried Einstein so much that he called it "spooky action at a distance." To understand what nonlocality is and what is really happening here, you need to delve into the very nature of reality, which we may never fully comprehend.

Nonlocality

In everyday life, we know that you need to touch an object to make it move. Only its immediate environment affects the object. Or, in order for an action in one point to affect another point, something in the space between these points must mediate this action. For example, controlling a toy helicopter from a remote control does not happen through magical influence, but through radio waves. This is the so-called principle of locality. That is, each object in the Universe has its own place, and these objects are separated from each other by oceans of space.

If you think about it, it seems that this is the only way it should be. That is why, in Newton's time, many were very concerned about his law of gravity. This law stated that apples fall and planets are held near the Sun because everything in the Universe attracts everything else. People were not worried about this, but because, according to Newton's idea, this force acts at a distance instantly. Raise your finger on Earth, and all the distant planets in the Universe will immediately shudder. Let a little, but it doesn't get any easier. The force of gravity jumps from the earth to the apple and from the finger to the planets, ignoring the empty space between them. The longer you think about it, the more terrifying it seems.

Einstein reassured everyone by demonstrating his theories of relativity that gravitational influence is limited by the speed of light. Can you remember your reaction when you first learned that nothing can move faster than light? I remember thinking it was strange and kind of out of thin air. Many people are annoyed that there is some kind of incomprehensible speed limit in the world we live in. And this, of course, is sad that the speed limit deprives us of the possibility of long-distance space travel. But something else is important: you wouldn't want to live in a world without this limitation.

If there were no speed limit, then various repulsive situations would occur. For example, the French mathematician Paul Painlevé described a case in which a star could fly out of a black hole at infinite speed. That is, such an accelerated star from any infinitely distant point in the Universe could instantly destroy our solar system, and we would not even have time to understand or notice it, or even somehow calculate such a situation.

In fact, it's even worse. According to the theory of relativity, when the speed of light is exceeded, cause-and-effect relationships can be violated. So, the known laws of physics say that a killer star could fly to us from the future. Infinite speed is not an intuitive thing, and it often erases the concept of space. As soon as you say the words "infinite speed," it becomes clear that something is wrong here. Infinitely fast motion hardly has the right to be called motion. An object that "moves" is already at its destination. So how can you say that it is moving there?

Imagine a situation where a ball from another galaxy can hit a ball in your yard and come back, spending zero units of time on all this. This situation would be completely non-intuitive. Either a situation where one ball simply magically influenced another, or a situation where there is actually no space between the two balls. Do you understand why entangled particles are at least an alarming bell? If you don't understand, then for physicists, for example, it is so important to preserve the concept of space and the absence of such magic in our world that they are ready to admit the existence of any other magic, just to explain this long-range action.

The hypothesis of superdeterminism, which we also considered in previous sections, is that exactly how each experimenter in each laboratory in the world will conduct measurements was planned in advance. That is, at the moment of the creation of the Universe, all the initial conditions were laid in its fundamental structure, including a detailed schedule of each measurement, each detector, each experimenter. The entire Universe was programmed to give the appropriate results and create the illusion of an instantaneous connection between entangled particles. Such an explanation, of course, is extremely inconvenient and requires the recognition that we all act according to a pre-written script, like actors in a grandiose cosmic play.

Superdeterminism

Superdeterminism is the concept that everything in the Universe, including every experiment and every measurement, was predetermined at the moment of the Big Bang. It seems to the experimenter that he is free to measure photons at any angle and at any time, wherever he wants. But in fact, all his actions are strictly programmed so as to register particles in such a way that they look consistent, although there is no consistency in reality.

That is, for example, so that the experimenter does not conduct an experiment that is undesirable for the Universe, his nose may start to itch, or his wife may call him, etc. You may ask, what is this paranoid delusion? However, this hypothesis is supported, for example, by the Nobel laureate in physics and one of the founders of the Standard Model, Gerard 't Hooft. He believes that locality is so vital that physicists should consider even crazy-sounding ideas to preserve it. And that without locality, the basic laws of physics would be very difficult or even impossible to formulate.

't Hooft argues that some new law of physics might be able to reconcile the properties of particles with the ways people choose to measure them. What seems like a conspiracy today may be the result of a conservation law that we don't yet know about. (Remember this, it will be very useful to us later.)

Particles as Crystal Balls

One equally insane way to preserve locality is the assumption that particles are able to see the future and that particles can be influenced by events that, from our point of view, should happen in the future. According to this hypothesis, the future must be able to influence the present in the same way that the past does. Particles can be born already having a memory of what will happen. In particular, they can remember the settings of the polarizers they will later encounter and be prepared to react accordingly.

This idea has already been taken seriously, for example, by physicists Richard Feynman and John Wheeler, who clearly do not need an introduction. That is, yes, from the point of view of scientists, these options are much better than the destruction of space. And if the problem were only in entangled particles...

The Bubble Paradox

Turn on the light bulb. The atoms in the filament begin to emit photons. How do you imagine this process? Imagine the very first photon that flew out of the lamp. From the point of view of the layman, mechanics says that the direction of the photon's departure is not determined by any known law of physics. The photon from your lamp seems to fly in all directions at the same time, forming a bubble growing in space. And only when the bubble reaches some object, it bursts with a certain probability, concentrating all the energy of the bubble in one specific place.

Physicists call this the collapse of the wave function. You see the light from the lamp because many such bubbles burst on the retina of your eye. This applies not only to the light from your lamp but also to any other light source, such as distant stars or galaxies.

If you don't see the problem yet, then one of the most distant objects that can be seen with the naked eye is the Andromeda galaxy, which is about 2.5 million light-years away from us. Now think about what happens when you look at this galaxy. Bubbles that began to spread 2.5 million years ago (people did not even walk on two legs then) reached a diameter of 5 million light-years, collapse on the retina of your eye, and do it instantly. Parts of the bubble, separated from each other by 5 million light-years, instantly learn that they need to stop spreading further, as if space does not matter to them.

This is the so-called bubble paradox. Again, someone will say that these photons are a trifle, and quantum mechanics is about the microcosm. But photons are the most common particles in the Universe described by the Standard Model, and as far as people can judge today, quantum mechanics is not a theory of the microcosm, it is a theory of the world, period.

Everything that exists is made up of the smallest particles. So that you can somehow assess the scale of the problem, Einstein, in an attempt to take a break from thinking about such behavior of light, you know what he did? He created the general theory of relativity.

Nonlocality Everywhere

Physicists are discovering more and more suspiciously mysterious nonlocal phenomena. They may all seem completely unrelated and distant from each other, but scientists say that's the point: they are connected at a deeper level. They may seem unworthy of attention and very far from our everyday experience, but let's not forget that a few drops of water can suggest the existence of an ocean, and looking at a falling apple can lead to the conclusion of the possibility of black holes. So rest assured: all examples of nonlocality, like pieces of a puzzle, will fit very organically into the madness that we will talk about a little later.

For example, when looking at the night sky, it seems to us that there is nothing unusual in it. But this only seems to be the case until you find out that matter in the early Universe could be distributed in so many different ways that it would be not just unlikely, but almost impossible for it to acquire the same density and the same temperature at all points. Any two galaxies or two large clusters of gas at opposite ends of our sky, at the very edge of the observable part of the Universe, are so far apart that light from the time of the Big Bang has not yet had time to fly from one galaxy to another. That is, you understand, they don't even see each other, they could not exchange energy or matter in any way, and yet they are very similar. American physicist Charles Misner said: "It is extremely difficult to explain why the sky is not dotted with spots." Observations have shown the consistency of objects that have never had the physical ability to interact with each other. And in 1972, the Russian theorist Yakov Zeldovich dared to suggest that a certain type of quantum nonlocality could explain the homogeneity of the cosmos. He dared because, I remind you, to say that locality is violated here is to say that space does not fulfill its functions. And if nonlocality really exists in nature, then it will destroy any science, because the basis of the scientific method is the identification of causes and the prediction of consequences.

But how do you establish cause-and-effect relationships if objects can magically influence each other instantly and at any distance? If something calls locality into question, it also calls space into question, and therefore it calls space-based theories into question. And this, for a second, is any theory that we have.

Einstein understood that the principle of locality, and with it our understanding of space, could be wrong. A few months before his death, Einstein reflected on what the disappearance of space could mean for our understanding of the world. "Then nothing will be left of my castle in the air, including the theory of gravity, as well as all of modern physics," said Albert Einstein. Even Niels Bohr, who disagreed with Einstein on many other issues, called long-range action irrational and completely incomprehensible.

Meanwhile, physicists who study black holes believe that matter in these cosmic vacuum cleaners can jump from one place to another without overcoming the distance between them. But, as Mayer writes, the main mystery is not there, but in the core of black holes - in the singularity. Where do you think the singularity is in a black hole? The general theory of relativity says that the matter inside reaches infinite density and space-time is torn apart, like an overloaded bag.

And the question "Where is the singularity?" implies the presence of space. How can we ask "where" if the space relative to which the position of the singularity should be determined no longer exists? We literally can no longer say "over there" or "here" or "15 meters to the right." A paradox, and therefore the answer also sounds paradoxical: in a black hole, the singularity does not exist anywhere, and at the same time, it exists everywhere. This is not easy to comment on.

As we can see, spatial anomalies crawl out from everywhere: in experiments in the quantum field, in the paradoxes of black holes, in the large-scale structure of the Universe. In all these examples, physics enters the twilight zone. Distance can lose its meaning. The universe becomes unrecognizable, appearing in different contexts. They have a striking similarity, which suggests that physicists are feeling different parts of the same elephant.

The Holographic Principle

"We believe that there is a three-dimensional world that exists even when no one is looking at it, and that it contains real objects, such as apples and waterfalls." - Donald Hoffman

Black holes are chilling objects that would be better off not existing. If you don't think so, then you simply never seriously imagined them. It would seem that so many strange things have been said about them, we have just talked about the incomprehensible location of the singularity inside. Well, what else can you add? But no, they continue to amaze. In general, Jacob Bekenstein and Stephen Hawking calculated that black holes increase their size in an extremely suspicious way, atypical for the three-dimensional world.

Imagine that you have a box that can fit one item. If you take another box, the length of the edge of which will be twice as large, then the surface area will be four times larger, and the volume will increase eight times. That is, if you can cram one object into the first box, then eight of the same objects will fit into the second. This is the so-called square-cube law, which Galileo demonstrated 400 years ago. This is how geometry works in the three-dimensional world. Can you imagine it working any other way?

But the fact is that this does not apply to black holes at all. Well, that is, look, from our point of view, it would be normal if everything were like with a box. That is, if doubling the radius of a black hole increased the area of its sphere by four times, and the volume and, accordingly, the capacity by eight times. However, this does not happen. Let's go slowly: when a black hole doubles in radius, the area of its sphere increases, as expected, by four times, but its volume does not increase eight times, as expected, but also by four. That is, it is as if in the example with the second box we visually got space for eight objects, but we could only cram four, despite all the seeming volume of space inside.

"Something would prevent you from shoving the fifth item in there. This is possible only in one case: in fact, increasing the width and length of the hole increases its capacity, but the additional height does not give anything, as if this measurement is illusory." - George Musser.

That is, the object black hole looks three-dimensional, but behaves like a two-dimensional one. What is this? A two-dimensional object in three-dimensional space?

And here's the catch. Black holes are not small, intangible particles. They can easily engulf the entire solar system, but they are very far away from us, so you might think that this strange behavior of theirs does not concern us in any way. However, this story has very far-reaching consequences. Hawking and Bekenstein quickly realized that this rule applies not only to black holes but to all other spaces. If you do not understand how this is possible, then Donald Hoffman explains it with a simple example: the maximum amount of information that six spheres can contain will be greater than the maximum amount of information that one large sphere can contain, in which these six could fit. That is, the volume literally does not play any role, only the surface area is important.

In our ordinary world, far from black holes, objects also look three-dimensional but behave like two-dimensional ones. I want you to understand very well what is meant. If you try to cram exactly as many things as a specific area of space visually suggests, then this area of space will collapse into a black hole, which will already take up as much space as it needs. This is called the holographic principle. Physicists Leonard Susskind and Gerard 't Hooft were engaged in its study. Susskind says: "Here is the conclusion that 't Hooft and I came to: the three-dimensional world of our ordinary experience, the universe filled with galaxies, stars, planets, houses, stones, and people, is a hologram, an image of reality encoded on a distant two-dimensional surface." This new law of physics, called the holographic principle, states that everything inside a certain region of space can be described using bits of information located on its boundary.

The Holographic Principle and the AdS/CFT Correspondence

The holographic principle and the AdS/CFT correspondence are important concepts in modern theoretical physics, offering profound and sometimes counterintuitive insights into the nature of space, time, and reality.

The holographic principle, proposed by Leonard Susskind and Gerard 't Hooft, states that all the information contained within a certain volume of space can be described on its surface. The idea originated from the study of black holes. As Stephen Hawking showed, information about the material absorbed by a black hole can be encoded on its event horizon, which led to the suggestion that three-dimensional space can be described on a two-dimensional surface.

This principle has far-reaching implications for our understanding of the universe. It suggests that our three-dimensional world may be a hologram, that is, a projection of two-dimensional information.

The AdS/CFT correspondence (Anti-de Sitter/Conformal Field Theory), proposed by Juan Maldacena, is a specific realization of the holographic principle. It establishes a connection between the theory of gravity in (d+1)-dimensional Anti-de Sitter space (AdS) and conformal field theory (CFT) in d-dimensional space. This correspondence suggests that theories in different dimensions are equivalent and that gravitational processes in the bulk AdS space can be described without gravity on its boundary, using field theory.

In simpler terms, imagine that we have two different theories: one is the theory of gravity, which describes how objects attract each other in space, and the other is conformal field theory, which describes the movement of particles and other physical processes. Juan Maldacena proposed the idea that these two different theories could be related. Specifically, he suggested that the theory of gravity in a certain space called "Anti-de Sitter space" could be related to conformal field theory in a space with fewer dimensions.

This correspondence, known as AdS/CFT, means that it is possible to describe gravitational phenomena in the space of gravity without using gravity itself. Instead, conformal field theory is used in a space with fewer dimensions.

Example with Black Holes and AdS/CFT

Black holes are central objects in understanding the holographic principle. Suppose there is a black hole with radius R. In ordinary three-dimensional geometry, the volume of this black hole should grow as

R^3, but the holographic principle states that information about this black hole should be encoded on its surface, the area of which grows as R^2. This means that the maximum amount of information that can be stored in a black hole grows with the square of the radius, not the cube, which corresponds to a three-dimensional volume.

Thus, the concept of space and time may not be what we are used to perceiving them, and may depend on more fundamental physical principles that we are just beginning to understand.

David Bohm's Holographic Universe

While working on the book, I came across the scientist David Bohm, whom I had not previously noticed in physics literature for some reason. When I got acquainted with his works more closely, I found out that he was one of the most outstanding physicists of the 20th century, collaborating on par with Albert Einstein himself. Bohm proposed his popular interpretation of quantum mechanics called the "pilot wave," which is very interesting and significant in the scientific community.

In addition, one of the most intriguing concepts that Bohm developed is the theory of the holographic universe. According to this model, reality is more plastic and changeable than we are used to thinking. All information about the whole is contained in each of its parts, as in a hologram. This means that every particle in the universe can contain all the information about the entire universe.

Karl Pribram, a renowned neurophysiologist, suggested that our brain functions like a hologram. This means that it processes information through interference patterns, similar to how holograms are created and interpreted using light.

Imagine that the brain processes information not linearly, like a computer, but like a hologram. In holography, information about the whole object is distributed throughout the hologram, and each of its fragments contains information about the entire object. This may explain why people are able to memorize huge amounts of information and instantly reproduce it.

This concept gave a new impetus to research in the field of neuroscience. It helped to understand the complex processes of memory, awareness, and perception. Pribram believed that the brain uses wave patterns to process and store information, similar to how waves on the surface of water create complex patterns.

Michael Talbot, author of "The Holographic Universe," discusses the possibility of explaining extrasensory phenomena through the holographic model. Talbot suggests that all parts of the brain are interconnected with the universe, which may explain phenomena that go beyond ordinary sensory perceptions. For example, extrasensory perception, premonition, and even telepathy may have their roots in the holographic nature of reality.

These experiences may indicate that consciousness is not limited to the physical body and can exist independently of it. This is of course just an assumption, but it sounds interesting.

Molyneux's Problem

Molyneux's Problem was first formulated in 1688 by the English natural philosopher William Molyneux in a letter to John Locke. The essence of the problem is as follows: would a person born blind, who received sight in adulthood, be able to immediately distinguish a cube and a sphere only by sight, without using touch?

Locke and Molyneux concluded that such a person would not be able to distinguish a cube and a sphere by sight alone. They believed that experience and learning are necessary to establish a connection between tactile and visual perceptions.

George Berkeley, in his work "An Essay Towards a New Theory of Vision" (1709), also supported this idea, noting that the connection between the world of touch and the world of sight is not natural, but is established only through experience.

Nowadays, this problem can be investigated experimentally. For example, between 2007 and 2010, Indian scientist Palan Singh led a study involving five patients who received sight after surgical treatment for cataracts. They were given a specially designed test within 48 hours of surgery.

The results showed that the patients could not immediately associate tactile knowledge of shape with visual perception. Their results were no better than random guessing. Only over time, through learning and experience, did they begin to better recognize objects, but still not 100%.

This experimental data supports the idea that the connection between different sensory systems is not innate but is formed through experience. Our senses, such as sight and touch, provide different types of information about the world around us, and only through the integration of this information with experience can we create a holistic view of objects.

Molyneux's problem challenges our ideas about perception and knowledge. How do we know what we know? Why do we believe that what we feel by touch should correspond to what we see? These questions have profound philosophical and psychological implications.

Everything from Bit

Physicists Niels Bohr and Werner Heisenberg, with their revolutionary theories, made a great contribution to our understanding of quantum mechanics. Bohr argued that "nothing exists until it is measured," and Heisenberg added that "what we observe is not nature itself, but nature exposed to our method of questioning."

In 2022, Anton Zeilinger and his colleagues received the Nobel Prize for experiments that dealt a serious blow to the concept of realism. Their work showed that our world is largely determined by information, not matter.

Francis Crick, one of the discoverers of the structure of DNA, in correspondence with Donald Hoffman wrote that, following Kant, one should distinguish the "thing-in-itself," which is fundamentally unknowable. In the end, Donald Hoffman himself believes that the Universe as we know it does not really exist until someone looks at it. But the point is not that the observer literally creates this Universe or that he can influence the world with the power of thought and other similar things, to which science is very skeptical.

Remember the double-slit experiment. What happens when we pass photons through two slits? If photons behaved like bullets, then on the screen behind the partition with two slits we would get a distribution in two stripes. But since photons, until they are fixed, propagate in all possible directions, they behave like probability waves, and on the screen we get a normal probability distribution. Most of the photons are registered in the middle, since their registration is most likely there, and the minority is on the sides, where it is least likely.

However, if we install detectors near both slits to find out which slit the photon passed through, the wave function collapses not on the screen behind, but near the first or second slit, since the measurement took place right there, and already measured photons on the screen behind will show a distribution in two stripes. If only one detector is installed near only one of the slits, the photon, even without being fixed in this slit, will still behave as if it was fixed in the neighboring one, and the pattern on the rear screen will be the same two-strip as in the case with two detectors.

Thus, for the collapse of the wave function, it is not enough just to measure, but even a simple fundamental possibility of obtaining information about the location of the particle. And this applies to any particle, since such phenomena in experiments with two slits demonstrate not only massless particles. In 2019, a world record was set under the leadership of Markus Arndt: it was possible to observe wave quantum properties in a huge molecule consisting of 2000 atoms. A little more, and scientists will be able to conduct similar experiments with viruses, which are sometimes considered one of the forms of life.

Therefore, quantum strangeness is not limited to the subatomic level, it extends to all matter. One of the most influential theoretical physicists of the 20th century, John Wheeler, said that "no entity can be considered fundamental if it does not translate all the physics of the continuum into the language of bits." He insisted that spacetime and its objects are not fundamental. Instead, he proposed the principle of "it from bit." In other words, matter from information. The principle is that information is fundamental, not matter. Matter arises from bits of information.

The Interface Theory of Perception

Donald Hoffman proposes a new perspective on our understanding of reality through what he calls the "interface theory of perception." It argues that we don't know what the true Universe is, but our perception is a kind of code for fitness that helps us survive and function in our environment.

Hoffman draws an analogy with the use of a computer. Imagine that you are writing a letter on a computer and saving it to your desktop. You see the file icon - a blue rectangle located in the center of the desktop. But this does not mean that the file itself is a blue rectangle and is located in the center of your computer. The color and shape of the icon are not real characteristics of the file, and its location does not correspond to the actual location of the file in the computer's memory. The file is stored as a set of bits of information, and the location of these bits has nothing to do with the icon on the desktop.

The icon does not try to convey the true nature of the file; on the contrary, its purpose is to hide this nature and save the user from unnecessary technical details. If you had to manipulate bits and electrical circuits instead of simply clicking on an icon, you would spend much more time and effort on tasks.

Computer interfaces have evolved to hide the complexity of the inner workings of a computer, and our senses do the same. Everything we see and feel is the Homo sapiens user interface. Space-time is our desktop, and physical objects, such as spoons and stars, are interface icons.

When you ask if the Moon really exists and whether we see its true color, size, shape, and location, it's like asking if the brush icon exists in Paint before you select it for drawing, and whether this icon reflects the true color, size, shape, and location of the brush inside the computer. The interface theory of perception suggests that our perception of objects was not formed to reflect objective reality, but to communicate the only thing that matters for evolution - information about fitness.

For example, a big scary bear is just an icon. But why not play with it? The fact is that evolution has not created an icon for ionizing radiation in our interface, so we do not feel the millions of particles that damage our body every day. For evolution, this is not important, as it does not prevent us from growing up and having children. But if you see a radiation hazard sign, you take it seriously, even though it has nothing to do with the radiation itself. Similarly, we should not touch the snake icon in our interface for the same reason we would avoid a torpedo on a submarine screen.

Evolution has shaped our senses to save our lives, so it's better to take icons seriously. But seriously doesn't mean literally. If I see a snake crawling towards me, I should take it seriously, but that doesn't mean there's something brown, smooth, and sharp-toothed when no one is looking.

Donald Hoffman also views our perception of space and time as a code for the energy costs of obtaining resources. For example, if it takes one calorie to get an apple, then it is perceived as being at a certain distance. Recent experiments confirm this idea: people who drink glucose drinks estimate the distance less than those who drink drinks with artificial sweeteners. More trained people also estimate the distance less than less trained people.

Critics, such as Michael Shermer, acknowledge that the interface theory of perception deserves serious consideration but express doubts about its limitations. Hoffman responds that science and technology allow us to manage our world better and better, but this does not mean that we understand its true nature. Just as Minecraft players become more and more proficient at manipulating its worlds, they don't necessarily understand the complex algorithms behind the game.

In Donald Hoffman's reflections on conscious realism, it is stated that consciousness is a manifestation of the mathematical nature of reality. According to this concept, consciousness is not part of the natural-science picture of the world, since the world we are exploring is our interface of perception.

Consciousness, according to Hoffman, does not fit into the framework of natural sciences, since it cannot be explained from the point of view of physical processes. For example, attempts to reduce subjective experience to the activity of neurons in the brain do not provide a complete answer to the question of the nature of consciousness.

Hoffman's theory of conscious realism proposes to look at consciousness as a manifestation of a more fundamental level of reality, which can be described mathematically and is not limited to physical phenomena.

CHAPTER 7: MATHEMATICAL REALITY

Cosmology and Magic

When we talk about real science, there is an unsolved mystery at its very heart. In December 1998, Max Tegmark, a renowned cosmologist, received an email that caused him excitement. It was a letter from a famous professor who criticized his articles:

"Dear Max, your crazy articles are not doing you any good. By submitting them to prestigious journals and failing to get them published, you are only amusing yourself. As an editor of a leading journal, I would never let your article through. You must understand that if you do not separate this activity from your serious research, you may jeopardize your future."

When Tegmark forwarded this letter to his father, he replied with a quote from Dante: "Go your own way and let people say what they want." Tegmark did just that, and today he is one of the most famous popularizers of science, a professor at the Massachusetts Institute of Technology, the author of numerous books, and a participant in many educational programs.

But what did he write that so angered the author of the letter? It's simple: Tegmark openly expressed his views on what he thinks our Universe is.

Neuropsychology and the Magic of the Mind

In his international bestseller "The Man Who Mistook His Wife for a Hat," Oliver Sacks describes 24 stories of people with mental disorders. Of all these interesting stories, one story stands out, which concerns two twins - John and Michael.

In 1966, Oliver Sacks met these twenty-year-old twins, who had been diagnosed with various diagnoses since childhood, from psychosis and autism to severe mental retardation. Most doctors considered them to be scientific idiots, savants, whose talent was limited to endless memory and the ability to instantly determine what day of the week any date falls on.

Here is one of the examples that Sacks describes. Once a box of matches fell off the table, and its contents scattered on the floor. The twins simultaneously shouted: "111!" And then John whispered: "37," and Michael repeated this number. John repeated it a third time and stopped. Sacks tried to count the matches and found that there were indeed 111.

Sacks asked them how they were able to count the matches so quickly. In response, he heard: "We didn't count, we just saw that there were 111 of them."

Impressed, he continued the conversation: "And why did you whisper 37 and repeat it three times?" The twins answered in unison: "37, 37, 37 - 111." Their answer was mysterious and incomprehensible, as was their ability to instantly determine the number of matches without counting.

Sacks describes how one day he caught the twins in a strange game: they exchanged six-digit numbers. Every time one called a number, the other nodded and happily answered with another six-digit number. Sacks wrote down these numbers and checked them against the tables at home. He found that all the numbers the twins exchanged were prime.

A prime number is a number that is divisible only by one and itself. For example, 7, 11, 13 are prime numbers. When it comes to small numbers, it is easy to determine which of them are prime and which are not. But when the number becomes six digits, this task becomes more difficult. However, the twins exchanged such numbers as if it were a common thing.

The next day, Sacks decided to conduct an experiment. He approached the twins and named an eight-digit prime number. The twins froze in deep concentration, and after half a minute, both smiled at the same time - they checked and realized that the number was prime.

After that, they began to exchange twelve, and then twenty-digit numbers. Sacks could not check these numbers because his tables were designed for a maximum of ten digits. In the 60s, only the most powerful computers could perform such a check, and even for them, it was difficult. There is no direct way to calculate prime numbers of this order at all, but the twins did it.

Sacks writes: "They see the arithmetic universe directly. Do we have the right to call it a pathology?"

The Level I Multiverse

Are there alien civilizations? The answer is very simple and unambiguous: yes. However, according to Max Tegmark's calculations, to get to the nearest such civilization, you will have to overcome at least a billion billion kilometers. Although with the same probability, aliens can be a billion billion times further away. This is not a very useful calculation, but the main thing is that they are definitely there, and here's why.

At the time of the release of this book, the officially confirmed oldest person on the planet among the living is Maria Branias Morera, who was born on March 4, 1907. During her schooling, she was probably told that the entire cosmos consists only of the solar system and a cloud of stars around it. But within the framework of her life alone, mankind's ideas about the size of the Universe have changed so much that the Universe, as Maria knew it in her school years, turned out to be only one of several hundred billion other universes that we can now observe and call galaxies.

Throughout the history of mankind, such an expansion of horizons has occurred repeatedly. Today we know that space is at least a billion trillion times larger than the greatest distances known to ancient hunters and gatherers. Moreover, Max Tegmark argues that according to the most popular cosmological model to date, the theory of inflation, space is not just huge, it is infinite.

In his words, the theory of eternal inflation is consistent with all modern observations and is the basis for most of the calculations and models presented at cosmological conferences.

What about aliens? Based on the fact that space is infinite and more or less evenly filled with matter, it can be argued that there is an infinite number of extraterrestrial life forms in space, even those that we cannot imagine. In infinite space, there is everything that is not prohibited by the laws of physics. What is this, a space snake? Obviously, there are snakes in space. There is literally everything in space. An infinite universe is a very strange place. For example, if the laws of physics allow the existence of some form of life that devours entire planets, then such monsters are guaranteed to exist somewhere.

Of course, we will not see this because of the limited speed of light and the expansion of the Universe. We live in the center of a bubble with a diameter of 93 billion light-years, beyond which space continues, but we cannot observe it.

Look at this model of the universe. It looks like a toy universe in which there are only four places for identical particles. This means that in this toy universe there can be only 16 possible combinations of matter. Now imagine that there are other toy universes around this toy universe. The question is: how often will the combinations of toy universes be repeated? Answer: we will need to check an average of only 16 neighboring universes to stumble upon a repeat.

Now transfer this example to our real observable universe. Of course, it has many more options for configuring matter, but these options are still limited. So, Tegmark says that according to a very conservative estimate, there are no more than 10^118 ways in which our observable universe can be arranged. Yes, this is a huge number: one followed by 10^118 zeros. This number is so large that if you turned all the matter in the observable universe into ink, you still wouldn't have enough to write it down completely.

And even despite this, this number is simply insignificant compared to infinity. And this means that if you look at the sky in any direction, then at a distance of approximately 10^10^118 diameters of the observable universe from you, at this very moment, your absolute copy will look at you, which has lived exactly the same life, thought exactly the same thoughts and did absolutely the same thing until the very last moment. Moreover, your double is on exactly the same planet, in exactly the same solar system, in exactly the same galaxy, and exactly the same observable universe.

The Boundary Between Physics and Metaphysics

"We take our theories too seriously and don't take them seriously enough." - Steven Weinberg, theoretical physicist, Nobel laureate.

Today we will consider strange ideas. Someone might say: "Why think about such metaphysical concepts?" But Max Tegmark argues that the boundary between physics and metaphysics is very unobvious and constantly shifting. For example, today we know that the Earth is shaped like a ball, but once it was a metaphysical hypothesis. Or the Earth's magnetic field, which we do not see - what is it if not metaphysics? Or slowing down time at high speeds, or particles that are in two places at the same time. What about the curvature of space? What about black holes? All this was once a metaphysical abyss, but today they are established facts of the physical world.

So, the boundary between physics and metaphysics is not determined by the strangeness of theories, as one might think, but only by the fundamental possibility of their experimental verification. And not even all physicists think so. It is becoming increasingly clear that theories based on modern physics can actually be predictive, empirically testable, and falsifiable.

There are as many as four levels of parallel universes, and for me personally, the most interesting question is not whether the multiverse exists, since the existence of its first level is beyond doubt, but how many levels are inside it. - Max Tegmark

But wait, we can't set up an experiment and check that space continues infinitely beyond our observable Universe? Tegmark says that we do not need to check this because the parallel universes formed by infinite space, and all the other parallel universes that we will talk about today, are not theories, but predictions of some theories.

Let me explain with an example. Einstein's theory gives an accurate prediction of how the planet Mercury moves. Can physicists test this? They can, and they check and find that the predictions of the theory are fulfilled with an accuracy at the limit of the measuring capabilities of the instruments. Further, the theory also predicts that light rays change their trajectory near massive objects due to the curvature of space. Arthur Eddington experimentally confirmed this in 1919. What else? Gravitational time dilation is also an experimentally confirmed fact. But the general theory of relativity also predicts things that we will probably never be able to test experimentally. For example, it describes to some extent the properties of space inside black holes. How do you check what's inside? You can, of course, fly into a black hole, but you will not be able to transmit observations outside for publication in a scientific journal.

And yet, all the predictions of the theory about the internal structure of black holes are taken very seriously by scientists, and no one dares to call them unscientific because other predictions of the theory work with great accuracy.

Tegmark writes: an important feature of physical theories is that if you like one of them, you will have to "buy" it in full. You can't say, "I like how general relativity explains Mercury's orbit, but I don't like black holes, so I want to do without them." You cannot "buy" general relativity without black holes. General relativity is a rigid mathematical construct that does not allow for fine-tuning. And you will either have to accept all its predictions or invent from scratch another mathematical theory that agrees with all the successful predictions of general relativity and at the same time predicts that black holes do not exist. This turns out to be an extremely difficult task, and so far such attempts have ended in nothing.

And according to the same principle, the theory of inflation has its own verified predictions. This is a very successful theory, and therefore it is necessary to take seriously those of its predictions that seem unverifiable, in particular, infinite space and parallel universes.

Even those of my colleagues who don't like the idea of the multiverse are now inclined to admit that the main arguments in its favor make sense. In general, criticism has changed from "This doesn't make sense and I hate it" to "I hate it." - Max Tegmark

The Strangeness in the Discovery of General Relativity

Physics reveals to us a reality far more complex than we could have imagined. Should we be surprised by this? No, evolution has endowed us with intuition only for those aspects of physics that were important for the survival of our distant ancestors. That's why we're shocked when liquid helium starts flowing upwards at low temperatures. But there are other phenomena that, although amazing, for some reason do not seem so to anyone. We don't even pay attention to them, as we didn't this time either.

The general theory of relativity is a rigid mathematical construct. Its predictions work with incredible accuracy. But no one is surprised at how the greatest theory in human history was discovered. Did Einstein look through a telescope and discover his theory? No. Perhaps he took some measurements to discover it? Also no. Instead of any experiments and observations, he sat at home for nine years and drew on paper.

These runes and pentagrams, which for most people on the planet do not mean anything at all, we call mathematics and do it with such a look, as if we understand what is at stake. Perhaps I will surprise someone, but the most knowledgeable people in mathematics, mathematicians, openly admit that in general, they have no idea what mathematics is. As the English philosopher Sir Michael Dummett once said: "The two most abstract scientific disciplines - mathematics and philosophy - cause the same bewilderment about what they actually do. Moreover, this bewilderment is caused not only by ignorance, it is difficult to answer this question even for specialists in the relevant fields."

But we'll think about it later. Right now, I'm trying to draw your attention to something else. Let's take the same gravity. Have you ever noticed the strangeness of objects falling under the influence of this very gravity?

Let's drop something and watch. For example, we have a falling ball. Here it falls and in one second overcomes a certain segment of the distance. What do you think, what segment of the distance will it overcome in the next second? I will not keep you in suspense - in the next second the ball will fly three times a larger segment, in the third second - five times more, in the fourth - seven times more, in the fifth - nine, then eleven, and so on.

Look again. What do you think is the strangeness of this sequence? If you look closely, you won't notice a single even number. The fall of any object is a sequence of odd numbers, which was discovered by Galileo. You can measure segments not once a second, but, for example, once every 5 seconds or once every 2 minutes - it doesn't matter. Regardless of the time interval you choose, you will always get just such a sequence. The ball falls as if the Universe knows exactly what odd and, therefore, even numbers are. This is a strict mathematical law, which, like any real law, has no exceptions. A law that is somehow woven into the fabric of the universe.

So amazing things often go unnoticed, and we take them for granted. But it is from such things that our understanding of physics and the Universe as a whole is formed.

Regularity vs. Chaos

If the ball fell a little differently each time, it would probably have stumped Einstein himself. Let me remind you of his words from a letter to the mathematician Maurice Solovine: "You find it strange that I speak of the knowability of the world as a miracle or an eternal mystery. Well, a priori one should expect a chaotic world that cannot be known through thinking."

Meanwhile, in a chaotic world, the brain probably simply could not have developed. For example, according to American neurobiologist Dean Buonomano, if someone had to formulate the whole essence of brain function in two words, the best definition would probably be "predicting the future." The brain is constantly doing mathematical calculations. For example, you most likely do not know why one of the faces in a pair seems more attractive to you than the other. But your brain knows. He's already figured it all out.

All these calculations of facial attractiveness are so complex that there is a channel on English-language YouTube dedicated only to researching which faces the human brain considers attractive, and there are already more than five hundred videos. That is, literally, the attractiveness of a face is about specific numbers, percentages, ratios, and proportions that you may not even suspect, but your brain has always known them. It performs mathematical calculations and predicts that there will be good offspring with this person. We call it attractiveness or beauty.

It is clear that this applies not only to the face. For example, many girls go to the gym and diligently squat with a barbell to build volume in the gluteal muscles. But, as it turns out, their volume is not as important as the bend of the lower back. Men rate the most attractive angle at 45.5 degrees. You ask why these numbers and ratios? Some of this can be explained, but at the same time, there are many specific numbers in the world, the origin of which is not clear. It is not known why the number pi, that is, the ratio of the circumference of a circle to its diameter, is found in various branches of physics, and it is not clear why it is so.

However, the number pi has already surprised everyone, and physicists are used to it as something completely natural. However, there are other strange numbers. For example, Feigenbaum's constant. Mitchell Feigenbaum worked at the famous Los Alamos Laboratory, which, among other things, was engaged in the development of the atomic bomb. One day he was given a cool modern HP 65 pocket calculator, which, adjusted for inflation, cost almost $5,000. Feigenbaum was fascinated by the new toy and, studying the behavior of one simple function on it, found that the sequence of numbers that he received as a result of calculations approaches a certain number.

When Feigenbaum investigated other equations, he found that this mysterious number also appears there. He concluded that he had discovered a certain universal pattern that somehow marks the transition from order to chaos. Although he could not find an explanation for this. At first, physicists were skeptical about this because it was hard to believe that the same number could characterize the behavior of different systems. His first article was peer-reviewed for six months and eventually rejected. However, very soon experiments showed that many things behave according to Feigenbaum's predictions. Its constant arises when measuring the dynamics of populations of living beings, the reaction of the eye to flickering light, atrial fibrillation, and the behavior of water droplets in a faulty faucet. Now this number is called the Feigenbaum constant, and it is known in the scientific world.

The Mysticism of Mathematics

Nobel laureate Eugene Wigner once said a phrase that later went viral: "The incredible effectiveness of mathematics in the natural sciences is something bordering on the mystical, as there is no rational explanation for this fact." Have you ever thought about what you do when you listen to music? I mean, what is it to listen to music and why do we get so much pleasure from it?

In school, we were told about the Pythagorean theorem and about Pythagoras himself. But what we were not told at school was that Pythagoras was the founder of a totalitarian sect named after himself. Its adherents worshiped numbers and believed that mathematics was literally God. Their motto was "Everything is number." To give you an idea of how serious it was there, when one of the students, Hippasus, mathematically proved that not all things can be expressed in integers, after a while he was found drowned.

So, Pythagoras discovered that music is mathematical and that the most pleasing to the human ear are specific ratios of vibrating strings, namely two to one (2:1), three to two (3:2), and four to three (4:3). These combinations of keys became the basis of classical music, most folk music, as well as pop and rock music. Thus, Pythagoras discovered that the harmony of sounds that we feel reflects the relationships that take place in a seemingly completely different world - the world of numbers.

I don't know how many times variations of this question will be repeated today, but how is this possible? The German mathematician Gottfried Leibniz wrote on this subject: "The pleasure we get from music comes from calculations, but unconscious calculations. Music is nothing but unconscious arithmetic." Arthur Schopenhauer believed that everything that exists is the embodiment of the world will, and music is its most direct manifestation. "Music, unlike other arts, is an imprint of the will itself. That is why its effect is so much stronger and deeper than the effect of other arts, because the latter speak of the shadow, while music speaks of the essence."

Thanks to the repeatedly confirmed truth of Leibniz's statement, music is nothing more than a way to directly and really comprehend those large numbers and numerical relationships that we can generally know only indirectly in concepts. And here's what's interesting: people with acquired or congenital savant syndrome, like the twins I described at the beginning, often have superpowers not only in mathematics but also in the same music. This suggests, as Oliver Sacks says: random numbers, and indeed any arbitrariness, did not bring any pleasure to the twins. They looked for meaning in numbers, probably in the same way that musicians look for harmony in sounds.

Oliver Sacks noted that in prime numbers, which the twins liked so much, it turns out that there really is some mystical hidden pattern that was accidentally discovered in 1963 by the mathematician Stanislav Ulam, and which even we, ordinary people, can see. Ulam was sitting at a very long and very boring lecture, trying to entertain himself somehow. He began to draw vertical and horizontal lines on a sheet of paper to start composing chess studies, but instead began to number the cells. He put one in the center, and then, moving in a spiral, two, three, and so on. At the same time, he mechanically noted prime numbers. It turned out that prime numbers line up in a certain harmonious pattern.

Surprised, Ulam returned from the lecture and created a computer visualization of what 90 million prime numbers would look like, and saw this. This is what is now called the "Ulam spiral." Why do numbers that are divisible without a remainder only by themselves and by one give such beauty?

Level II Multiverse

Alan Guth is a physicist and cosmologist who proposed the idea of cosmic inflation, which predicts the existence of a Level I multiverse. But it turns out that it also predicts the existence of a Level II multiverse, as demonstrated by Alan Guth, Andrei Linde, Alexander Vilenkin, and other physicists.

Once, Guth, in his report read at the Massachusetts Institute of Technology, noted that if we discover some object in nature, then the scientific approach suggests that we must also find the mechanism that generated this object. For example, cars are built in car factories, rabbits are born with the participation of rabbit parents, and planetary systems are born during the gravitational collapse of giant molecular clouds. Therefore, we must assume that our entire Universe was generated by a mechanism for creating universes. And here's what's important: car factories, rabbits, and giant clouds of dust produce many copies of what they create. A universe that contains only one car, one rabbit, and one planetary system does not seem natural.

According to this logic, the mechanism that gave rise to our universe must have given rise to many others. The Level I multiverse is simply one universe with infinite space, where sooner or later everything repeats itself. But the Level II multiverse is already a more interesting structure.

In physics, there are nine fundamental particles called fermions. Each of them has its own mass, and these masses are very different from each other. But what is interesting is that if you look at these masses, they look like they were chosen randomly.

Imagine that you are throwing nine darts at a dartboard. Each dart hits a random spot, and the distance from the center of the target to each dart will be different. Similarly, the masses of fermions look random, as if they were "scattered" on the mass scale without any pattern.

This is strange because we are used to thinking that everything in the Universe has its own reasons and patterns. But the masses of fundamental particles do not seem to obey any rules. This raises an important question for scientists: why are the masses of particles what they are? Is there any hidden meaning in this, or is it just a coincidence?

But let's go further. Imagine that you need to adjust the round knob that is responsible for the density of dark energy. Dark energy is a repulsive force in the Universe, so you can't overdo it, otherwise, stars and galaxies will not be able to form in space. But at the same time, if you do not tighten it, the Universe will very quickly collapse under the influence of gravity. You ask, what is the range of settings in this case? Physicists have calculated that the maximum possible value is about 10 to the 120th power of kilograms per cubic meter, and the minimum value is 10 to the minus 97th power of kilograms per cubic meter.

So, what do you think, with what accuracy do you need to turn the knob so that our Universe can exist? Answer - the angle of rotation must be set with an accuracy of more than 120 digits after the decimal point. It turns out that no matter how you twist it, you won't be able to hit it accurately. And yet, obviously, some mechanism did it for our universe.

And the Universe has many such "pens." Max Tegmark writes that the scientific community is gradually beginning to understand that many of them are very finely tuned. For example, if the electromagnetic forces were weakened by about 4%, the sun would immediately explode. How can this be explained? There can be three options here. The first is a chain of lucky coincidences. However, the scientific method does not tolerate unjustified coincidences. As Tegmark writes, to say that "my theory requires an unreasonable coincidence to agree with observations" is the same as saying, "my theory is wrong."

The second option is God, divine intervention. However, this option is not much better than the previous one, since it does not explain anything and itself raises a huge number of other questions.

And the third option is the theory of inflation. It assumes the presence of space, which is infinitely expanding. In other words, it "boils," and "bubbles" appear in this space, like in a pot of boiling water. Each bubble is a Level I multiverse with infinite space inside. And all these endless bubbles together form a Level II multiverse.

If you have a question about how infinite space can be enclosed in the finite volume of these bubbles, then I will tell you even more: for an outside observer, all these universes can look like formations smaller than an atom, which probably look like this - a black hole of subatomic universes, their space is endless.

Thus, what we call the Big Bang was not the beginning, but rather the end - the end of inflation in our region of space. In other areas, inflation usually continues forever. Needless to say, most of the Level II parallel universes are dead due to failed settings?

When talking about the Level II multiverse, Tegmark often appeals to the statistical approach. And his predictions are in excellent agreement with the data. And if you think about it, it's absurd in its own way. How can accidents have a pattern? It sounds like an oxymoron.

Regularity vs. Chaos 2

Belgian mathematician Adolphe Quetelet conducted a large-scale study of various parameters of the human body. He measured, for example, the chest circumference of 5,738 Scottish soldiers and the height of 100,000 French recruits. Expressing all the readings graphically, Quetelet obtained a bell-shaped curve, which we now call the normal distribution curve. The more data he had on a certain parameter, the clearer this curve became. For example, if we take a parameter such as height, then the absolute majority of people have approximately the same height, and deviations affect the minority: on the left of the graph, there will be very short people, and on the right - very tall people.

Quetelet also built similar curves for moral qualities, such as the propensity for crime, intellectual abilities, and so on. To his surprise, he found that all human characteristics obey this same normal curve.

But what is truly amazing is that Quetelet discovered this curve, known to astronomers from astronomical observations, back in the middle of the 15th century. How can it be that astronomical, biological, and social processes are connected by some universal law? The mere fact that the distribution of a wide variety of properties obeys the same normal curve is remarkable in itself. But this is not enough. Even the distribution of the average level of successful serves in the major baseball league and the profitability of stock indices obey a normal distribution.

Moreover, if the distribution deviates from the normal curve, it should usually be carefully checked. For example, if the distribution of grades in English in a school differs from the normal one, this suggests checking the grading rules adopted there.

Mathematical patterns can be traced in a wide variety of areas. In 1906, researcher Francis Galton, a second cousin of Charles Darwin, made an important observation at a country fair. Visitors were asked to guess the exact weight of a slaughtered bull. 787 people took part in the competition. Among them were both farmers who understand this and people far from cattle breeding. After the fair, Galton calculated that the average of all answers was 1,197.5 pounds (about 547.5 kg). How close do you think this number was to the actual weight of the bull? The error was less than 1%. Absolutely chaotic answers from different participants led in aggregate to a very accurate result. This phenomenon has been repeatedly reproduced in various fields and has been called the "wisdom of the crowd."

This effect underlies such phenomena as democracy, where decisions are made based on the votes of a large number of people, as well as services such as Wikipedia or the online platform "Kulu," created in 2015 by a group of scientists. On this platform, people can make their predictions about certain events, and the platform shows the average voting result. Many of the predictions made came true with high accuracy.

Can mathematical patterns really permeate everything? Many studies and observations show that even in randomness there is a certain order that can be described mathematically. These patterns help us better understand the world and even predict future events with some accuracy.

Ramanujan's Genius

In January 1913, a talented Cambridge mathematician named Godfrey Harold Hardy received a package of documents with a cover letter. The author of the letter, Srinivasa Ramanujan, stated that he had made striking progress in mathematics and asked Hardy to publish his works, since he himself did not have the funds for this. Attached to the letter were 11 pages of technical results from various branches of mathematics, most of which were already known mathematical theorems, but some Hardy saw for the first time. Hardy immediately realized that these formulas could only be derived by a mathematician of the highest class, and they must be true, since no one could have invented them.

Srinivasa Ramanujan was a young Indian who had no formal mathematical education and never attended university. Hardy and his colleague John Littlewood were convinced that they were dealing with a genius who had single-handedly walked the century-long path of European mathematicians. Hardy helped Ramanujan move to Cambridge to work together.

The problem was that until now no one understands the method by which Ramanujan derived his formulas. Hardy said that Ramanujan's ideas about mathematical proof were very vague. Ramanujan impromptu issued complex arithmetic theorems, the proof of which would require modern computers. He claimed that his formulas were drawn to him in a dream by the goddess Namagiri.

Ramanujan left three volumes of notes containing extremely strong theorems without any comments or proofs. In 1976, another 130 pages of his notes were found for the last year of his life, containing 600 formulas without proofs. Almost all of them were subsequently proven. Mathematician Richard Askey said that Ramanujan's work in the last year of his life is comparable to what some great mathematician could have done in a lifetime.

The work of deciphering his last diary was extremely difficult. Mathematician Bruce Berndt said that the discovery of this manuscript caused a stir in the mathematical world, similar to the discovery of Beethoven's Tenth Symphony. Physicist and mathematician Stephen Wolfram wrote that Ramanujan's complex formulas hid a story behind them. Many of his results seem like random facts from mathematics, but their work in recent decades shows that they obey mathematical laws.

Freeman Dyson said that Ramanujan had some magic tricks that we don't understand. Ramanujan's story is reminiscent of the scale of his genius. A film about him, "The Man Who Knew Infinity," was even made in 2015.

His notebooks, which contained brief summaries of his results, were studied for decades after his death as a source of new mathematical ideas. And the most fantastic thing is that his formulas are used today in string theory and to study black holes, although such terms as string theory and black hole did not exist during his lifetime. Ramanujan somehow answered the questions of theoretical physics that no one had even asked yet.

One way to explain this might be that the brain evolved to see a certain pattern in the world, some kind of mathematical one. Perhaps his neurons took over the function of calculation in the way that the brain calculates the proportions of the face. Presumably, his neurons were involved in calculating mathematics. However, most likely, the first option is more correct, or both are correct.

If we consider the first option, then Ramanujan described theorems that are now used in string theory. And string theory is an explanation of the world at the most fundamental level.

The Incredible Effectiveness of Mathematics in Physics

Galileo Galilei once said: "The great book, I mean the universe, which is always open to our eyes, is written in the language of mathematics, and its signs are triangles, circles, and other geometric figures." Galileo emphasized that without knowledge of mathematics, we cannot understand nature. This statement remains true to this day, as mathematics surprisingly finds application in physics, revealing the secrets of the universe to us.

If we approach the question of the effectiveness of mathematics in the natural sciences from an everyday point of view, we might think that people observed the physical world and understood some properties of addition, subtraction, and so on. For example, if you have three apples and you eat one, you will have two left. It can also be assumed that any person will sooner or later come to the conclusion that space has three dimensions. From this point of view, it is not surprising that mathematics and physics are closely related.

But the main problem with this logic is that mathematics is successfully used in areas that are as far from human perception as possible. Take Einstein, for example. Many people think that he received the Nobel Prize for the theory of relativity, but this is not so. The Nobel Committee stubbornly refused to recognize his candidacy for decades, despite the fact that he was nominated by such prominent scientists as Lorentz, Planck, and Bohr. Why?

Various reasons are given, including the lack of experimental data. Everything he did, all his work was complex mathematics, without any experiments. Therefore, some members of the Nobel Committee did not understand the essence of his theory, and on the other hand, they were very skeptical that the slowing down of time and the curvature of space were something real. It's hard to blame them for this, as it seemed incredible.

This continued until the incoming experimental data could no longer be ignored. But even then, the Nobel Committee, paralyzed by indecision, awarded Einstein the Nobel Prize not for the theory of relativity, but for what is considered his least significant achievement - the explanation of the photoelectric effect.

So why does mathematics describe so well what a person has never encountered in the entire history of his existence? Why, for example, is the unattainable world of subatomic particles so well described by mathematics learned by counting vegetables? And why do the formulas of Ramanujan, a man who had nothing to do with physics at all, find their application in the most modern physical concepts after 100 years? Why, after all, even Ramanujan's mentor, the same Godfrey Hardy, who was literally proud that his works contained nothing but pure mathematics, and in his famous book "A Mathematician's Apology" wrote: "I have never done anything useful; none of my discoveries have either directly or indirectly, increased or decreased the good or evil in the world," why even his formulas have found their application in reality, for example, in the Hardy-Weinberg law, the fundamental principle on which geneticists rely in the study of population evolution?

Level III Multiverse

On the night of September 26, 1983, near Moscow, alarms sounded at the command center of the early nuclear warning system. The computer reported that intercontinental ballistic missiles had been launched from the territory of the United States of America. The level of reliability of the readings was maximum. In the minds of everyone who was at that moment in the command center, only one thought surfaced - the Third World War. That night, Lieutenant Colonel Stanislav Yevgrafovich Petrov was on duty. His heart was pounding and his breath caught. "I could not get up from the chair, my legs were taken away," he recalled.

According to the charter, Petrov was obliged to report the attack by launching a chain of orders, which would lead to a corresponding nuclear strike on the United States. "I had only a few minutes to inform the country's leadership about the threat. The missiles were supposed to explode on our territory in just half an hour." The tens of thousands of nuclear bombs accumulated over the years of the arms race were about to fulfill their purpose. Most of them were not even atomic, but hydrogen. For those who do not know what a hydrogen bomb is: in a hydrogen bomb, an atomic bomb acts as a trigger for a reaction.

"It seemed to me that my head turned into a computer. A lot of data, but they did not form a single whole." No one knows if the leadership of the Soviet Union would have initiated a retaliatory strike if Lieutenant Colonel Petrov had reported the attack. This was quite likely, as the situation was very tense then. It was the height of the Cold War. Reagan was no longer shy in expressions and called the USSR "the evil empire" and "the focus of evil in the modern world." And three weeks before the incident, the leadership of the USSR gave a paranoid order to destroy a civilian airliner flying from New York to Seoul. The liner, due to a pilot error, went off course and flew into the airspace of the USSR, where it was shot down by our interceptor. As a result, 269 people died, including US Congressman Larry MacDonald.

The situation was such that both the US and the Soviet Union were seriously considering options for preemptive nuclear strikes against each other because each country was afraid that the other would do it first. The chances were 50/50. And now think about it: the fate of the whole world at that moment depended on whether a single calcium atom would or would not get into a specific synapse of the prefrontal cortex of the brain of Lieutenant Colonel Petrov, causing the excitation of a specific neuron and sending an electrical signal by it, which would launch a cascade of activity of other neurons jointly encoding the thought "false alarm."

"I picked up the phone and reported to the duty officer that the information coming from my command post was false. The computer crashed." It remained only to wait until the missiles, if they were indeed launched, invaded the airspace of the USSR and were detected by radars. This was supposed to happen in 18 minutes, but it didn't. The next two days after the experienced shock, according to the lieutenant colonel's son, his father slept. Six months later, it turns out that the failure occurred due to the fact that the sun's rays were reflected in a certain way from the clouds directly above the base and blinded the satellite.

After the collapse of the Soviet Union, the whole world learns about this story. Stanislav Petrov will receive the prestigious German Media Award for his contribution to the public good. In New York, at the UN headquarters, he will be presented with a crystal statuette with the inscription "To the man who prevented a nuclear war." Petrov will become a laureate of the Dresden Prize, awarded for the prevention of armed conflicts, and will star in a documentary about those events along with Kevin Costner.

This is the story we know, but there is another reality. The calcium atom that triggered the cascade of events in the lieutenant colonel's brain is a microscopic object subject to the laws of quantum mechanics. Therefore, an atom can be in two slightly different positions. According to the many-worlds interpretation of quantum mechanics, on the night of September 26, the universe split into two realities. Right now, in parallel with our world, there is another one where the calcium atom did not hit the right synapse in the lieutenant colonel's brain, and Petrov made the opposite decision - he reported an attack, and a nuclear war began. One can only guess what this world looks like now. This is a kind of Schrödinger's cat experiment, but on the scale of an entire planet.

The Many-Worlds Interpretation of Quantum Mechanics by Hugh Everett does not need an introduction. It has been talked about a lot, and many physicists in recent decades have moved from ignoring or mocking it to seriously considering the possibility of an infinite splitting of the universe. Max Tegmark cites an informal but revealing poll of physicists in 1997, where the majority opted for the classic Copenhagen interpretation rather than parallel universes. But already in 2010, in Harvard, no one voted for the Copenhagen interpretation at all, and the absolute majority recognized the correctness of the many-worlds interpretation.

Konrad Lorenz said that important scientific discoveries go through three phases: first, they are ignored, then fiercely attacked, and finally, dismissed as well-known. Judging by the survey data, after passing the first phase in the 1960s, Everett's parallel universes are now between the second and third phases.

Max Tegmark notes that many do not understand how one can split into copies of oneself and not notice it, always feeling like the same person. You can try to understand this and accept it only with the help of a thought experiment. There is no law of physics that would prohibit creating your full copy with all your memories. Imagine that you fell into a deep sleep, and after that, you were cloned using the super technology of the future. If, after waking up, you are not told which of you is a clone, then you will never be able to be sure that you are the original. You will both come from the same past and feel like you have lived a long life, even though one of you only appeared yesterday.

If the universes of the first and second levels are within the framework of traditional cosmology, then the multiverse of the third level, as imagined in the many-worlds interpretation of quantum mechanics, suggests something completely different. Here, every possible version of history, every decision, every possible outcome is real and exists in parallel with each other.

The story of Lieutenant Colonel Petrov, in which he made the opposite decision and reported an attack, illustrates this concept. In the many-worlds interpretation of quantum mechanics, each possible outcome of this incident is realized in a separate branch of reality. And although for us, living in a certain branch, this may seem like something abstract or fantastic, within the framework of the many-worlds interpretation, this is a natural feature of the quantum world.

What is the Foundation of Reality Made Of?

The foundation of physical reality consists of mathematical objects such as Hilbert spaces and wave functions. Hilbert space is a mathematical structure that is used to describe the properties of quantum systems and their states. It includes infinite-dimensional vector spaces, which are used to formalize quantum mechanics.

The wave function is a mathematical object that describes the state of a quantum system and allows you to predict the probabilities of various measurement results. It is the basis for calculating probabilities, amplitudes, and other quantities in quantum mechanics.

Thus, mathematics acts as the basis of physical reality in the context of quantum theory, where Hilbert spaces and wave functions help describe the behavior of microscopic particles and systems.

Frank Wilczek, in his publication for the online edition, writes: "In my scientific career, there have been many different experiences, some of them leading me to unusual states of consciousness. But I had only one experience that could be described as mystical. I was there alone, inside a metal box the size of an airplane hangar, looking down at the equipment that people use to experimentally study the fundamentals of nature. And then it happened. It intuitively occurred to me that the complex calculations I had done with pen and paper could somehow describe this completely different realm of existence, namely the physical world of particles, tracks, and electrons created by the mechanism I was looking at. There was no need to choose, as is often the case with philosophers, between mind or matter. It was mind and matter together. How could this be? Why should it be so? And yet I somehow suddenly realized that it could and should be so. The great mystery of the correspondence of mathematical language to the laws of physics is an amazing gift that we are unable to understand and that we may not deserve."

This experience revealed to him the amazing gift that mathematical language gives to the laws of physics. He felt that the laws of nature could be understood through the language of mathematics, and that this language is not just an abstract concept, but also reflects the deep structures of reality. It was a sense of the unity of mind and matter that became a mystical experience for Wilczek.

Level IV Multiverse

In the previous section, we discussed in detail why what a person directly perceives with his senses cannot be objective reality. I mainly covered the radical point of view of cognitive psychologist Donald Hoffman. But what few people will argue with is that the picture of the world we perceive is extremely subjective. We see only a somewhat distorted model built by our brain.

The 19th-century physicist and ophthalmologist Hermann von Helmholtz described the mechanism of this phenomenon, summarizing that we are not looking at reality, but at a model of reality created by our brain. The model of the world is our inner reality. What external reality really looks like outside of our senses is a big question.

However, we found that we have access to external reality through mathematics. Your perception tells you that you are looking at a solid stone, but its mathematical description shows that the stone is mostly made up of empty space between constantly vibrating particles. We trust the mathematical description more than subjective feelings, otherwise, we would not have built modern civilization with its technologies.

Why is external reality described by mathematics? This question has tormented people for millennia and today is more acute than ever. Is mathematics an invention or a discovery? Can we say that mathematics exists independently of the human mind? Are we discovering mathematical truths like new islands and continents, or is mathematics just a human invention, a tool?

The question of the nature of mathematics is closely related to the question of the existence of God. Mathematics and physics are often seen as two different disciplines. However, Max Tegmark proposes the idea that our entire physical world is a giant mathematical object. The problem of the effectiveness of mathematics arises only when we consider them to be different disciplines. If they are one and the same, everything falls into place.

Platonism and Reality

The belief that mathematical objects exist in reality and are more real than what we see goes back to Plato. Platonism argues that mathematical forms do not exist in the same way as ordinary physical objects. They have no spatial location and do not exist in time.

Max Tegmark believes that all structures are equivalent, and therefore mathematical structures are reality. Subatomic particles are not solid objects, but only clusters of mathematical properties. The very space of our physical world is a purely mathematical object.

The Level IV multiverse is different realities corresponding to different fundamental laws of physics and governed by different mathematical equations. If at the lowest level reality is a mathematical structure, then its parts consist of relationships between mathematical blocks, not their properties.

Perhaps the most surprising thing is that the Universe, despite its complexity, can be described by a simple mathematical formula. As in the case of the Mandelbrot set, which is described by the formula $Z = Z^2 + C$, the complexity of the Universe can be the result of such simple mathematical expressions.

The question of how humanity fits into this mathematical picture of the world remains open. Perhaps we are part of a larger mathematical structure that manifests itself through the laws of physics, and our understanding of this will help us better understand ourselves and the Universe in which we live.

What is a Human Being According to Max Tegmark?

In his mathematical universe hypothesis, Max Tegmark views a person as a complex mathematical pattern in the space-time continuum.

According to Tegmark, our consciousness and perception of the world are the result of the interaction of complex information processes in the brain. These processes allow our brain to create models of the world and ourselves, interacting with them.

Tegmark's main theses about human nature include the following aspects:

- **Consciousness and Matter:** Tegmark acknowledges that it is not yet clear how exactly physical matter gives rise to consciousness. However, he considers the possibility of creating a theory of consciousness in the future that is as holistic and convincing as the theory of electromagnetism.
- **Mathematical Connection:** Tegmark points out that consciousness has some mysterious mechanism for accessing the mathematical world. This mechanism either discovers, creates, or formulates a wealth of abstract mathematical forms and concepts.
- **Evolution of Mathematical Abilities:** He notes that even animals have basic mathematical abilities, and these abilities are innate and develop under the pressure of natural selection. However, human mathematical abilities far exceed the skills necessary for survival.
- **Mathematics and the Physical World:** Tegmark wonders how mathematical laws describe the physical world with such accuracy and why these laws have such complexity and beauty.
- **Four-Dimensional Space-Time:** According to the theory of relativity, every point in the past, present, and future exists in reality, and therefore objects such as the Earth and the Moon form unchanging patterns in space-time. Human space-time patterns are the most complex in the observable Universe.
- **Quantum Mechanics:** Tegmark also considers the influence of quantum mechanics, where each of us can branch into many branches, forming a beautiful pattern in an infinite mathematical universe.
- **Consciousness as Information Processing:** According to Tegmark, consciousness is the way information feels that it is being processed by certain complex methods. It occurs when the model of yourself in your brain interacts with the model of the world in the same brain or with itself.

Max Tegmark's mathematical universe hypothesis, which states that physical reality is a mathematical structure, faces challenges in terms of its verification and falsification.

Falsification of the Hypothesis

Tegmark notes that the hypothesis can be considered falsified if physicists, even without having a complete description of physical reality, stop finding mathematical patterns in nature. In other words, if it turns out that physical laws and phenomena do not lend themselves to mathematical description, this will be a serious argument against his hypothesis.

Tegmark's hypothesis of multiverses has its critics who put forward strong arguments against it. In particular, critics point to the difficulties of empirical testing of the hypothesis, the lack of observational evidence, and other theoretical problems. Tegmark responds to this criticism, acknowledging that all these statements are valid, but he still believes in the truth of his hypothesis and is ready to risk all his property by betting on the existence of multiverses.

Tegmark emphasizes that mathematics is a great mystery that we have yet to unravel. He points out that in the history of different peoples, magic was a complex system of knowledge that gave adherents special abilities. If you replace the word "magic" with "mathematics," then his statement about the occult level of reality that can be conquered will remain relevant.

Tegmark's hypothesis, although it meets with considerable criticism, remains an interesting concept that raises discussions about the nature of reality and the role of mathematics in its description. It encourages scientists and philosophers to think about the deep connection between mathematical structures and physical reality, even if its final verification remains a difficult task.

The Journey Continues: An Invitation to Explore Further

Thank you for joining me on this exciting journey through the labyrinths of reality and consciousness. We explored the amazing world of quantum physics, delved into the mysteries of evolution and consciousness, and tried to understand the complex questions that have troubled humanity for centuries.

We saw that reality can be much more complex and amazing than we are used to perceiving it. We realized that consciousness is not just a product of the brain, but something more, something that goes beyond our understanding.

We pondered the questions "What am I?" and "What is happening around us?". These questions lead us into the unknown, opening up new horizons of knowledge.

But the search for answers does not end there. On the contrary, it is only beginning. Every new discovery, every new idea brings us closer to understanding the mystery of consciousness and reality.

We can doubt everything, but we cannot doubt our own existence, our own consciousness. This is the only unshakable truth from which we can start in our search.

Do we live in a simulation? Is consciousness a fundamental property of reality? Will we ever be able to fully unravel the mystery of consciousness?

So far, there are no definitive answers to these questions. But it is the search for these answers that makes our life meaningful and interesting. It encourages us to explore, reflect, and develop.

If you are interested in diving deeper into the world of quantum physics and its connection to consciousness, I recommend reading my next book "The Quantum Code: Deciphering the Secrets of the Universe". In it, we will take a closer look at quantum biology, various quantum interpretations, and the concept of quantum consciousness.

I hope this journey has been informative and inspiring for you. I wish you success in your further search for truth!

"X" - @woodson1900

davidwoodson84@gmail.com

www.ingramcontent.com/pod-product-compliance
Lightning Source LLC
Chambersburg PA
CBHW071058240526
45471CB00016B/2088